STRAIN-ENGINEERED
MOSFETs

C. K. Maiti
T. K. Maiti

CRC Press
Taylor & Francis Group
Boca Raton   London   New York

CRC Press is an imprint of the
Taylor & Francis Group, an **informa** business

CRC Press
Taylor & Francis Group
6000 Broken Sound Parkway NW, Suite 300
Boca Raton, FL 33487-2742

First issued in paperback 2017

ISBN-13: 978-1-4665-0055-6 (hbk)
ISBN-13: 978-1-138-07560-3 (pbk)

Library of Congress Cataloging-in-Publication Data

Maiti, C. K.
    Strain-engineered MOSFETs / C.K. Maiti, T.K. Maiti.
        p. cm.
    Includes bibliographical references and index.
    ISBN 978-1-4665-0055-6 (hardback)
    1. Metal oxide semiconductor field-effect transistors--Reliability. 2. Integrated circuits--Fault tolerance. 3. Strains and stresses. I. Maiti, T. K. II. Title.

    TK7871.99.M44M248 2012
    621.3815'284--dc23   2012031209

Visit the Taylor & Francis Web site at
http://www.taylorandfrancis.com

and the CRC Press Web site at
http://www.crcpress.com

Contents

Preface

Microelectronics fabrication is facing serious challenges due to the introduction of new materials in manufacturing and fundamental limitations of nanoscale devices that result in increasing unpredictability in the characteristics of the devices. The downscaling of complementary metal-oxide-semiconductor (CMOS) technologies has brought about increased variability of key parameters affecting the performance of integrated circuits. In silicon-based microelectronics, technology computer-aided design (TCAD) is well established not only in the design phase but also in the manufacturing process. Device design procedures are now more challenging due to high-performance specifications, fast design cycles, and high yield requirements. Design for manufacturability and statistical design techniques are being employed to meet the challenges and difficulties of manufacturing of nanoscale-integrated circuits in CMOS technologies.

As mainstream CMOS technology is scaled below the 22 nm technology node, development of a rigorous physical and predictive compact model for circuit simulation that covers geometry, bias, temperature, DC, AC, radio frequency (RF), and noise characteristics becomes a major challenge. While introducing new device structures, innovation has always been an important part in device scaling and the integration of new materials. It is envisioned that the right combination of global biaxial and local uniaxial strain could provide additional mobility improvements at low electric fields. Written from an engineering application standpoint, the book provides the background and physical insight needed to understand new and future developments in the modelling and design of n- and p-MOSFETs at nanoscale.

Understanding predictive modelling principles to gain insight in future technology trends is important for future circuit design research and integrated circuit (IC) development. Technology CAD is a bridge between the design world and the manufacturing world. Compact models are useful not only for long-term product design but also for early evaluation of a technology for circuit manufacturing. The ultimate goal of predictive technology and process compact modelling is to describe any process technology accurately. The concepts of process compact and process technology modelling are essential to achieve the necessary knowledge transfer, which has proven to be useful in the silicon manufacturing world.

The focus of this book is on state-of-the-art MOSFETs, implemented in high-mobility substrates such as Ge, SiGe, strained Si, and ultra-thin germanium-on-insulator platforms, combined with high-k insulators and metal-gate. The book consists of 10 main chapters covering substrate-induced strain engineering in CMOS technology, process-induced stress, electronic properties of strain-engineered semiconductors, strain-engineered MOSFETs, noise in strain-engineered devices, technology CAD and reliability of strain-engineered MOSFETs, process compact modelling, and process-aware design of strain-engineered MOSFETs, and looks beyond the 22 nm node.

Several excellent books and monographs have appeared on multigate MOSFETs, high-mobility substrates, and Ge microelectronics and strained semiconductor physics. Numerous papers have appeared on strained Si and process-induced strain, but there is a lack of a single text that combines both the strain-engineered MOSFETs and their modelling using technology computer-aided design. We attempt to summarise some of the latest efforts to reveal the advantages that strain has brought in the development of strain-engineered MOSFETs. We have included important works as well as our own research and ideas by the research community, and due to space limitations, we have referred to only representative papers and listed books recently published in related areas for additional reading.

The book is mainly meant for final-year undergraduate and postgraduate students, scientists, and engineers involved in research and development of high-performance MOSFET devices and circuits. We hope this book will help in process technology development and design of strain-engineered MOSFETs. It may also serve as a reference book on strain-engineered heterostructure MOSFETs for active researchers in this field.

We thank Chhandak Mukherjee for contributing to Chapter 6.

C. K. Maiti and T. K. Maiti
Kharagpur, India

About the Authors

Dr. C. K. Maiti received his BSc (Honors in Physics), B.Tech. (in Applied Physics), and M.Tech. in radiophysics and electronics from the University of Calcutta; MSc (by research) from the University of Technology in Loughborough, UK; and PhD from the Indian Institute of Technology–Kharagpur in 1969, 1972, 1974, 1976, and 1984, respectively. Dr. Maiti joined the Department of Electronics and Electrical Communication Engineering of the Indian Institute of Technology–Kharagpur in 1984 as an assistant professor and was appointed professor in 1999. He is currently the head of the department and leads the semiconductor device/process (TCAD) simulation research group within the department. He has also contributed significantly in the areas of low-temperature dielectric formation on Si, III-V semiconductors, and group IV alloy layer films. He has published four books in the silicon-germanium and strained silicon area. Dr. Maiti has edited the *Selected Works of Professor Herbert Kroemer*, published by World Scientific (Singapore, 2008). He has also served as the guest editor for the *Special Issues on Silicon-Germanium of Solid-State Electronics* (November 2001) and *Heterostructure Silicon* (August 2004). He has authored/coauthored more than 250 technical articles and conference publications. Dr. Maiti's current research interests cover various aspects of semiconductor process and device simulation of heterojunction transistors involving strained layers.

Dr. T. K. Maiti obtained his MSc degree with a gold medal in physics from Vidyasagar University, India, in 2005. In November 2005, he joined Microelectronics Centre, Indian Institute of Technology–Kharagpur, India, to carry out research work on technology CAD (TCAD) of strain-engineered MOSFETs. He completed his PhD in engineering from the Department of Electronics and Telecommunication Engineering at Jadavpur University, India, in 2009. Dr. Maiti then moved to Canada to work at McMaster University as a postdoctoral fellow on design of advanced inorganic devices (i.e., silicon, III-V, and II-VI) on silicon via modelling of the device structure, fabrication process, and electrical performances in the Engineering Physics Department from February 2010 to February 2012. Since March 2012 he has been working as a researcher at HiSIM Research Centre, Hiroshima University, Japan. Dr. Maiti's current research interests include the experimental and simulation analysis of future generation semiconductor devices as well as the development of reliable compact models for circuit and system-level simulation.

List of Abbreviations

BEOL: Back end of line
BOX: Buried oxide
BPTM: Berkeley Predictive Technology Model
BTBT: Band-to-band tunneling
BTI: Bias temperature instability
CD: Critical dimension
CESL: Contact etch stop layer
CMOS: Complementary metal-oxide-semiconductor
CMP: Chemical mechanical polishing
CVD: Chemical vapour deposition
DFM: Design for manufacturability
DFT: Density functional theory
DFY: Design for yield
DG MOSFET: Double-gate MOSFET
DIBL: Drain-induced barrier lowering
DoE: Design of experiments
DQW: Double quantum well
DRIE: Deep reactive ion etching
DSL: Dual-stress liner
e-SiGe: Embedded SiGe
EDA: Electronic design automation
EOT: Equivalent oxide thickness
EPM: Empirical pseudopotential method
EUV: Extreme ultraviolet radiation
FCC: Face-centred cubic
FEA: Finite element analysis
FEOL: Front end of line
FET: Field-effect transistor
FN: Fowler–Nordheim
FUSI: Fully silicided
GAA: Gate-all-around
GAA MOSFET: Gate-all-around MOSFET
GeOI: Germanium-on-insulator
GIDL: Gate-induced drain leakage
GR: Generation recombination
GSI: Giga-scale integration
GSMBE: Gas source molecular beam epitaxy
HBT: Heterojunction bipolar transistor
HCI: Hot-carrier injection
HDD: Highly doped drain

HFET: Heterostructure field-effect transistor
HH: Heavy hole
HK-MG: High-k/metal gate
HOT: Hybrid orientation technology
HP: High performance
IC: Integrated circuit
ITRS: International Technology Roadmap for Semiconductors
KOZ: Keep-out zone
LF: Low frequency
LH: Light hole
LNA: Low-noise amplifier
LOP: Low operating power
LRP: Limited reaction processing
LRPCVD: Limited reaction processing chemical vapour deposition
LSF: Least-squares fit
LSTP: Low standby power
MBE: Molecular beam epitaxy
MG: Metal gate
MOS: Metal-oxide-semiconductor
MOSFET: Metal-oxide-semiconductor field-effect transistor
MUGFET: Multigate MOSFET
NBTI: Negative bias temperature instability
NW: Nanowire
OMVPE: Organometallic vapour phase epitaxy
PB: Planar bulk
PBTI: Positive bias temperature instability
PCM: Process compact model
PDK: Predictive process design kit
PR: Piezoresistance
PSD: Power spectral density
PSS: Process-induced strain
PTM: Predictive technology model
QW: Quantum well
R-D: Reaction-diffusion
RBL: Relaxed buffer layer
RIE: Reactive ion etching
RPCVD: Reduced-pressure chemical vapour deposition
RTA: Rapid thermal annealing
RTCVD: Rapid thermal chemical vapour deposition
RTN: Random telegraph noise
RTS: Random telegraph signal
S/D: Source/drain
SCE: Short-channel effect
SEG: Selectively epitaxial growth
SEM: Scanning electron microscope

SEU: Single-event upset
SGOI: SiGe-on-insulator
SiGe: Silicon-germanium
SIMS: Secondary ion mass spectrometry
SMT: Stress memorisation technique
SMU: Source monitor unit
SNWT: Silicon nanowire transistor
SO: Spin-orbit
SOI: Silicon-on-insulator
SPCM: SPICE process compact model
SPT: Stress proximity technique
SQW: Single quantum well
SRB: Strain-relaxed buffer
SS: Subthreshold slope
SSDOI: Strained Si-directly-on-insulator
SSOI: Strained Si-on-insulator
STI: Shallow trench isolation
TCAD: Technology computer-aided design
TDDB: Time-dependent dielectric breakdown
TG MOSFET: Tri-gate MOSFET
TSV: Through-silicon via
UHVCVD: Ultra-high-vacuum chemical vapour deposition
ULSI: Ultra-large-scale integration
UTB: Ultra-thin body
VCO: Voltage-controlled oscillator
VLS: Vapour-liquid-solid
VLSI: Very-large-scale integration
VS: Virtual substrate
VW: Virtual wafer
VWF: Virtual wafer fabrication

List of Symbols

E: Energy
E_g: Band gap of semiconductor
E_C: Conduction band energy
E_V: Valence band energy
$\Delta E_C, \Delta E_V$: Strain-induced change in the energy of carrier subvalleys
N_C, N_V: Effective densities of states
n_i: Intrinsic carrier concentration
T_{Si}: Silicon channel thickness
t_{ox}, T_{ox}: Oxide thickness
k, k_B: Boltzmann constant
k_x, k_y, k_z: Wave vectors
W: Width of the transistor
H_{fin}: Fin height
W_{fin}: Fin width
L, Lg: Channel or gate length
V_{FB}: Flat-band voltage
V_T, V_{TH}, V_t, V_{th}: Threshold voltage
C_{ox}: Gate oxide capacitance
C_d: Depletion capacitance
Q_{ox}: Oxide charge density
ΦB: Energy difference of the Fermi level in the bulk region
V_{bi}: Built-in potential across the source/drain channel
Φ: Potential
Φ_0: Difference between the electron affinities of Si and SiO_2
ψ_s: Surface potential
μ: Mobility
μ_{eff}: Effective mobility
X_d: Channel depletion layer length
X_j: Source/drain junction depth
$Q_{is}, Q_s, Q_i, N_{inv}$: Inversion charge density
v: Injection velocity near source region
T: Temperature
C_L: Load capacitance
S: Subthreshold slope
f: Frequency
g_m/I_d: Transconductance-to-drain current ratio
g_m: Transconductance
V_{dd}, V_{DD}: Power supply voltage
R_{ds}: Drain/source resistance
I_{on}: On-state current

I_{off}: Off-state current

I_D, I_{DS}, I_d: Drain current of MOSFET

I_G: Gate tunneling current

V_{DS}, V_d: Drain/source voltage of MOSFET

V_{GS}: Gate/source voltage of MOSFET

R_C: Backscattering coefficient

σ: Stress

$\vec{\sigma}$: Stress tensor

F: Force

A: Area

$\sigma_{11}, \sigma_{22}, \sigma_{33}$: Normal stresses

$\sigma_{12}, \sigma_{23}, \sigma_{13}$: Shear stresses

ε: Strain

$\vec{\varepsilon}'$: Strain tensor

$\bar{S}$: Elasticity tensor

S_{11}, S_{12}, S_{44}: Parallel, perpendicular, and sheer components of elastic modulus

m^*: Conductivity effective mass

H: Hamiltonian

Ψ: Wave function

Ξ, ξ: Deformation potential

$\pi_{||}$: Longitudinal piezoresistance coefficient

$\pi_{\perp}$: Transverse piezoresistance coefficient

$\tau(\varepsilon)$: Strain relaxation time

μ_n^0, μ_P^0: Electron and hole mobility without the strain

m_{nl}, m_{nt}: Electron longitudinal and transfer masses in the subvalley

$m_{Pl}, m_{P,h}$: Hole light and heavy effective masses

F_n, F_p: Quasi-Fermi levels of electron and holes

$\psi_k(\vec{r}), \psi_k(\vec{r}, \varepsilon)$: Eigenfunctions for the unstrained and strained conditions

$S_I(f), S_V(f)$: Current and voltage noise power spectral density

ΔN: Fluctuation in the number of carriers

τ_c, τ_e: Mean capture and emission time constants

ΔI: RTS pulse amplitude

α_H: Hooge's parameter

$N_t, N_T(E_F)$: Trap density

γ: Frequency exponent of noise

S_{ID}/I_D^2: Normalised drain current noise

λ: Tunneling attenuation length

Z_t: Position of the trap measured from the Si/SiO$_2$ interface

v_{TH}: Thermal velocity of electrons

σ_n: Electron capture cross section of the traps

E_T: Trap energy level

$\Delta N_{t,eff}/N_{t,eff}$: Effective change in trap density due to stress

$\Delta \Phi_B(\sigma)$: Ground energy-level shifts in the inversion layer for applied different types of stresses

$f_t(\sigma)$: Stress-dependent trap occupation function

$S_{VG}(f)$: Gate voltage noise power spectral density
D_{it}: Interface state density
λ_{sc}: Scattering parameter in correlated mobility fluctuation model
S_{IB}: Base current noise power spectral density
I_B: Base current
V_{BE}: Base-emitter bias voltage
V_{CE}: Collector-emitter bias voltage
A_E: Emitter area
L_S: Screening length
ΔI_B: RTS amplitude in base current
h_{21}: Current gain
$s_{11}, s_{22}, s_{12}, s_{21}$: S-parameters
f_T: Cutoff frequency
E_{tao}: DIBL coefficient
N_{ch}: Channel doping concentration
T_{SiN}: Nitride cap layer thickness

1

Introduction

It is not the strongest of the species that survives, nor the most intelligent, but the one most responsive to changes.
—Charles Darwin

In the field of microelectronics, the planar silicon metal-oxide-semiconductor field-effect transistor (MOSFET) is perhaps the most important invention. It started in 1928 when J. E. Lilienfeld proposed the concept of field-effect conductivity modulation and the MOSFET. William Shockley, John Bardeen, and Walter Brattain invented the transistor in 1947, and with the discovery of silicon dioxide (SiO_2) passivation for the Si by Atalla in 1958, the Si MOSFET era started. Since then MOSFET performance has been improved at a dramatic rate via gate length scaling, and complementary metal-oxide-semiconductor (CMOS) is currently the dominant technology for integrated circuits. As the technology scales almost every 2 years, the transistor integration capacity doubles (Moore's law), gate delay reduces by 30%, energy per logic operation reduces by 65%, and power consumption reduces by 50%. Table 1.1 shows CMOS technology outlook extrapolated from the current International Technology Roadmap for Semiconductors (ITRS) trends. However, conventional CMOS scaling has now approached the fundamental limits, which include leakage in channel and gate, diminished bulk effect, transport in silicon, and increased power dissipation. The huge costs of scaling CMOS devices according to Moore's law have now left the silicon industry at a crossroad. As technology scales, the cost of a transistor goes down, but the cost of fabrication facilities, cost of mask set, and turnaround time increase for each generation. Lithographic challenges for future technology nodes have become a major concern. Implementation of extreme ultraviolet radiation (EUV) will help continue transistor size scaling. Although it will allow for increased device density, current scaling issues will be a major concern at smaller gate length devices. ITRS 2009 has projected scaling of the advanced MOSFETs covering the next 15 years through 2022. The evolution of the Si process technology after the 130 nm node is shown in Figure 1.1. Technology challenges for 10 nm CMOS and beyond will face process limitations such as patterning ultra-fine and random features, ultra-thin gate dielectric (~3 Å), and ultra-shallow junction (~3 nm). In the following, we shall address the recent developments, which have been the subject of a major research drive for the last 10 years aimed at finding new avenues to enhance the performance of MOSFETs.

TABLE 1.1

Technology Trend

High-Volume Manufacturing	2006	2008	2010	2012	2014	2016	2018
Technology node	65	45	32	22	16	11	8
Integration capacity (BT)	4	8	16	32	64	128	256
Delay = CV/I scaling	~0.7	>0.7	Delay scaling will slow down	Delay scaling will slow down	Delay scaling will slow down	Delay scaling will slow down	Delay scaling will slow down
Variability	Medium	Medium	High	High	Very high	Very high	Very high

With the 90 nm technology node, strain techniques have been introduced to efficiently increase the transistor drive current by enhancing the mobility of carriers in the channel. Stress has been incorporated in MOSFETs during CMOS processing. Process-induced stress is now a viable, competitive, and important key technology that will certainly be used to boost the performance of future technology generations. Intel has been in the

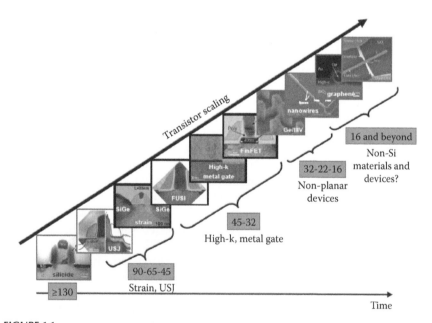

FIGURE 1.1

Evolution of the Si process technology after the 130 nm node. (After A. Shickova, Bias Temperature Instability Effects in Devices with Fully-Silicided Gate Stacks, Strained-Si, and Multiple-Gate Architectures, PhD thesis, Katholieke Universiteit-Leuven, 2008.)

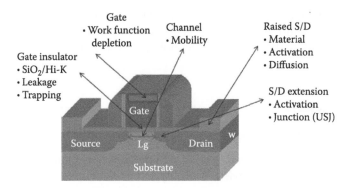

FIGURE 1.2
Main challenges for CMOS technology in 22 nm technology node. (After Maiti, T. K., Process-Induced Stress Engineering in Silicon CMOS Technology, PhD thesis, Jadavpur University, 2009.)

forefront in addressing the challenges by successfully driving transistor innovations from research phase to mainstream CMOS manufacturing. Process-induced stress engineering has kept the expectations high when Intel announced the tri-gate MOSFETs in 22 nm technology node in May 2011. The first 3D tri-gate transistor appearing at 22 nm will have a big impact on the industry. It has been predicted that the tri-gate FinFETs are both viable and capable of being tailored to suit the power/performance trade-offs for a range of applications and exceed the capabilities of present silicon-on-insulator (SOI) technology. Tri-gate FinFETs built on high-k/metal gate (HK-MG) technology, which will continue with Intel's 22 nm platform, have the flexibility for lower power consumption and much higher performance.

In scaling down CMOS technology beyond the 22 nm node, the semiconductor community will face further challenges. Figure 1.2 summarises some of the main challenges in the scaling of traditional planar bulk MOSFETs. As devices are scaled beyond the 22 nm node, various architectural and material changes in the traditional MOSFET would be required for efficient operation of the transistor. Innovative technologies such as new device architectures, mobility enhancement technology, high-k/metal gate dielectric integration, source/drain engineering, and enhanced quasi-ballistic transport channels may serve as possible solutions.

Technology development driven by Moore's law has so far played a vital role for the success of the semiconductor industry. In the last few years, the semiconductor industry has witnessed a quick development of a new area of micro- and nanoelectronics beyond the boundaries of Moore's law. The "more Moore" development is defined as a relentless scaling of digital functions and an attempt to further develop advanced CMOS technologies to reduce the cost per function. Today we have reached the end of classical Dennard scaling and are being confronted with a set of cumulative interrelated challenges at all levels, from system level down to atomic level, and require innovative processing steps and new materials. In 2005, the strategic research agenda and vision for "more than Moore" technology had been formulated in a systematic manner by the European Technology Platform for Nanoelectronics.

1.1 Technology Scaling

In order to improve the speed of ULSI/GSI devices, new materials and device structures are being proposed. Mobility enhancement techniques such as global (substrate) strain and process-induced (local) stress are currently the most promising for improving device performance. There are a number of ways to induce strain in silicon. Different types of strain have distinct effects on electron and hole mobilities. Starting from the 90 nm technology nodes, advanced CMOS technologies feature multiple process-induced stressors such as compressive and tensile overlayers, embedded SiGe, and multiple stress memorisation techniques. Large magnitudes of uniaxial channel are being incorporated in p-MOSFETs in the 65 nm technology node, and an even higher stress level is required beyond the 22 nm technology node. Local strain approaches are based on dedicated processing steps or process modules, such as shallow trench isolation, silicidation or metal gate electrodes, the use of liners and capping layers, dry etch processes, contact etch stop layers, and source/drain engineering. Various mobility enhancement technologies currently in use are shown in Figure 1.3. Although the terms *stress* and *strain* are used very often interchangeably, they have different meanings. Stress is the force per unit area that is applied to a given material, while strain is the material response to this external stress. The stress can be accommodated in the material by changing the interatomic distances or by material expansion/contraction by defect creation.

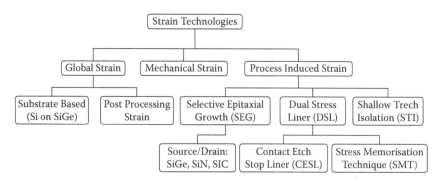

FIGURE 1.3
Different mobility enhancement technologies currently in use. (After Maiti, T. K., Process-Induced Stress Engineering in Silicon CMOS Technology, PhD thesis, Jadavpur University, 2009.)

1.2 Substrate-Induced Strain Engineering

To induce appropriate strain in the channel region of MOSFETs, various techniques have been introduced, such as substrate-induced strain, process-induced strain, and bending-induced strain. Optimisation of channel surface crystalline orientations for maximum carrier mobilities can also provide for a significant improvement in CMOS performance. Biaxial tensile silicon strain has long been known to increase electron mobility, but the strain-induced hole mobility increase is small at high vertical electric field. Substrate-induced strain engineering has become a critical feature in CMOS technology since it enhances the drain current without further gate length scaling. Recent progress has also demonstrated the evolution of the strained Si bulk MOS structure, such as the strained Si on SiGe-on-insulator (SGOI) MOSFET, and the strained Si-directly-on-insulator (SSDOI) MOSFETs. With a highly strained Si channel or a different orientation substrate in p-MOSFETs, the performance match between the n- and p-MOSFETs for CMOS applications might be achieved. Alternative channel materials with mobilities higher than silicon mobility, e.g., germanium or III-V semiconductors, can be used for device performance enhancement.

In Chapter 2, the issue of the substrates for strained-layer SiGe applications is addressed, followed by a short review of the present epitaxy techniques in use for SiGe research and production. A comprehensive review on state-of-the-art substrate-induced strain engineering methodologies in CMOS technology will be presented. Strain effect on various n- and p-channel MOSFETs in both inversion and accumulation regions are discussed. A systematic analysis of the strain effects on deeply scaled n- and p-MOSFETs with Si, SiGe, strained Si, strained Ge, and Ge channel is presented. Besides strained Si on the traditional (100) plane, it may be advantageous to change the crystal orientation to optimise CMOS circuit performance. Another way of enhancing channel mobility without the introduction of any new channel materials is the use of the hybrid crystal orientation technique. The carrier mobility of inversion layers depends on surface orientation and current flow directions, due to asymmetry of the carrier effective masses in the Si crystal lattice. Hybrid orientation technology (HOT) will also be discussed.

1.3 Process-Induced Stress Engineering

Process-induced (local) strain was first introduced into planar Si MOSFET transistors by Intel in 2002. Uniaxial strain is generated by local structural change near the channel region. The embedded SiGe (e-SiGe) under the source and drain regions resulted in larger than expected device performance enhancement, which is attributed to compressive channel. The strain is induced by the lattice mismatch between Si and SiGe. Owing to the relative ease of integrating process-induced strain modules in conventional CMOS processing, strain-enhanced scaling is now possible. However, uniaxial channel stress requires different stress types (compressive and tensile for n- and p-MOSFETs, respectively). Stress development in integrated circuits may occur at any stage of the manufacturing process from a variety of sources that affect the device performance. Several standard processing steps can be used to introduce uniaxial strain in silicon channel for MOSFET strain engineering. Various techniques have been proposed to incorporate strain in the channel region. Most successful among these has been the introduction of SiGe in the source/drain regions, use of tensile and compressive liners, as well as the stress memorisation technique. The two critical areas of stress development in integrated circuits are (1) front-end-of-line strain-engineered channel for increasing carrier mobility and (2) thermomechanical stress development near Cu through-silicon vias (TSVs) for 3D integration. A clear understanding of the evolution of stress/strain in integrated circuits and novel ways in which it can be characterised can lead to more effective strategies to mitigate or control the stress development. Equivalent scaling strategies such as strain-engineered MOSFET channels and 3D integration schemes are important for maintaining integrated circuits performance enhancement in future semiconductor technology nodes.

In Chapter 3, typical uniaxial technologies, such as embedded or raised SiGe or SiC source/drains, Ge preamorphisation source/drain extension technology, the stress memorisation technique, and tensile or comprehensive capping layers, stress liners, and contact etch stop layers, are discussed in detail. The importance of global and local strain techniques is outlined. Layout-dependent compact modelling of mobility, velocity, and threshold voltage in strain-engineered state-of-the-art transistors using e-SiGe, dual-stress liner, and shallow trench isolation stresses are discussed. Three-dimensional integration has emerged as a viable solution to achieve higher packing density. Toward 3D integration, through-silicon vias, which directly connect stacked structures die to die, are being employed. It is important to note that the through-silicon vias (TSVs) generate a stress-induced thermal mismatch between TSVs and the silicon bulk, which affects the performance of nearby transistors, diodes, and associated circuits. Thus it is important to study the impact of TSV-induced stress on device and circuit performance. TSVs also interact with polysilicon and shallow trench isolation layout pattern density. A summary of benefits of 3D integrated circuit (ICs) and key process steps involved in their fabrication, particularly relating to through-silicon vias, will be discussed. In nanometer-scale CMOS transistors performance variability is common and layout-dependent effects have become important. The important issues of device/circuit interactions for the 22 nm node will include discussions on variability; design for manufacturing and the impact of back-end technology elements on overall device performance will also be covered.

1.4 Electronic Properties of Strained Semiconductors

The band structure provides the information about the states of energy and the electronic dispersion relation under a specific condition. It is known that if the band structure of the material is modified, mechanical and electrical properties of the material will be also changed, such as effective mass and corresponding mobility. Band structure analysis provides details about strain effects on the electron/hole transport property. For instance, strain-induced lattice constant change will induce band warping in both the conduction band and the valence band. However, the effective mass change is much more important for holes in the valence band due to a strong correlation between six subbands. In Chapter 4, we shall briefly discuss the stress–strain relationships and their effects on the band structure, and a representative method of strain components in terms of elastic compliance constants is given. The basic physical definitions, such as the strain and stress tensors, are introduced. Different methods of calculating the effect of strain on the band structure are presented. The deformation potentials of the conduction and valence bands are calculated, and the band edge shifts and splitting are discussed in detail as strain effects. Since carrier mobility is a key parameter for the simulation of the electrical characteristics of semiconductor devices, several analytical models capable of capturing the dependence of mobility on temperature, doping, and electric field will be introduced. Various types of mobility models commonly used in simulation will be described in detail.

1.5 Strain-Engineered MOSFETs

As the MOSFET channel length enters the nanometer regime, short-channel effects (SCEs), such as threshold voltage roll-off and drain-induced barrier lowering (DIBL), become high, which hinders the scaling of planar bulk or silicon-on-insulator (SOI) MOSFETs. To overcome these problems, new device architectures as well as new gate stacks have been proposed. Multigate (also known as FinFET) devices are considered a promising architecture for replacement of conventional planar MOSFETs, offering a solution for overcoming the short-channel effects and providing better threshold voltage control at short gate lengths. In Chapter 5, different schemes of multigate devices are reviewed. The tri-gate devices will be the focus of this chapter because they are a good compromise between processing complexity and electrical performance. Although the gate-all-around (GAA) and the Π-gate structures show better electrical properties, they require more complex and costly processing for implementation. According to the ITRS, the strongest driver for high-k gate dielectrics comes from the need to extend battery life for wireless devices due to lower leakage currents, which include gate leakage, subthreshold leakage, and junction leakage. Strain engineering has become a critical feature now in CMOS technology since it enhances the drain current without further gate length scaling. Process integration issues such as power consumption, leakage current, metal gate electrodes, and high-k gate dielectrics will be covered in this chapter.

1.6 Noise in Strain-Engineered Devices

Among the different types of noise mechanisms present in semiconductors, low-frequency noise, typically observed to exhibit a dependence on frequency, is very important for analogue and mixed-signal applications. Low-frequency noise is known to degrade the spectral purity of nonlinear radio frequency and microwave circuits, such as oscillators and mixers, where the low-frequency, base band noise generates noise sidebands around the radio frequency (RF) or microwave carrier signal through up conversion into unwanted phase noise. In Chapter 6, fundamental noise sources in semiconductors are reviewed and their physical origins are analytically described. The low-frequency noise is discussed as a diagnostic tool for identifying traps and defects at the insulator/semiconductor interface, and as a device lifetime prediction tool for reliability analysis. The low-frequency noise behaviour in various strain-engineered devices such as strained Si MOSFETs, multigate FETs, FinFETs, silicon nanowire transistors, and heterojunction bipolar transistors will be discussed. We shall also discuss the strain effects on MOSFET operations such as threshold voltage, gate tunneling current, and low-frequency noise characteristics. For devices processed on strained Si, it is reported that the low-frequency noise increases when Ge from the SiGe buffer diffuses up into the active layer or when threading dislocations are present. For this purpose, low-frequency noise in strained Si MOSFETs is extensively studied.

1.7 Technology CAD of Strain-Engineered MOSFETs

Technology computer-aided design (TCAD) simulations allow one to explore new technologies and novel devices through physics-based modelling, optimise process and device performance, and control manufacturing processes through statistical modelling. All these are performed on a computer and are known as virtual wafer fabrication (VWF). Basic TCAD flow is shown in Figure 1.4. Technology modelling and simulation include the semiconductor

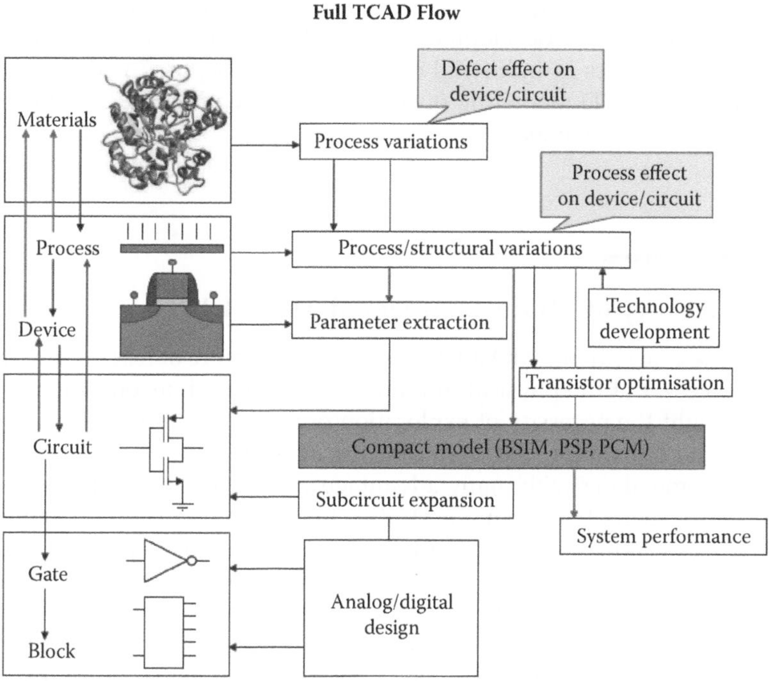

Full TCAD Flow

FIGURE 1.4
Compact multilevel technology/device/subsystem modelling flow. (After Maiti, T. K., Process-Induced Stress Engineering in Silicon CMOS Technology, PhD thesis, Jadavpur University, 2009.)

process modelling, and it is one of the few enabling methodologies that can reduce circuit development cycle time and cost. As the mainstream CMOS technology is scaled into the nanometer regime, development of a rigorous physical and predictive compact model for circuit simulation that covers geometry, bias, temperature, DC, AC, RF, and noise characteristics becomes a major challenge. Compact models have been at the heart of CAD tools for circuit design over the past decades, and are playing an ever increasingly important role in semiconductor manufacturing. Development of a compact model describing a new technology is essential prior to the adoption of the technology by the semiconductor industry. TCAD is currently being used for process and device design, manufacturing, and yield improvement.

Traditionally, a custom design is considered superior because it delivers higher performance and smaller die size, thus resulting in lower cost; however, this results in a longer design cycle (time to market) and is now a serious challenge for the 22 nm CMOS technology node and beyond. It is expected that the future of designs in 22 nm and beyond will be system design with design automation at all levels. Statistical fluctuations inherent in any IC manufacturing process cause variations in device and hence in circuit performance. Thus, product yield and manufacturing problems necessitate costly redesign cycles. Technology computer-aided design is an indispensable tool for development and optimisation of new generations of electronic devices in industrial environments. Chapter 7 is dedicated to the technology CAD modelling of strain-engineered MOSFETs in process-induced strain technologies.

1.8 Reliability of Strain-Engineered MOSFETs

Scaling the conventional MOSFETs has so far been more or less a straightforward process. But the physical limitations encountered beyond the 130 nm node brought the necessity of exploration of new gate stack high-k materials, mobility enhancers, and even new device architectures. The new technologies come along with many advantages, but also raise many concerns about their reliability. Systematic studies to determine the key parameters controlling the reliability are necessary. It is important to identify the intrinsic reliability problems of the advanced devices, to distinguish them from extrinsic effects of processing, and to suggest new methods for reliability improvement. In Chapter 8, the bias temperature instabilities (BTIs) of some of the new generation devices, such as high-k/metal gate (HK-MG) stacks, strain-engineered devices with enhanced mobility, and FinFET devices, are considered. Technology CAD has been used to study the effects of strain on the negative bias temperature instabilities (NBTIs) in process-induced strained Si p-MOSFETs and hot-carrier injection in process-induced strained Si n-MOSFETs.

1.9 Process Compact Modelling

Aggressive technology scaling has led to large uncertainties in device and interconnect characteristics for deep-submicron circuits. Many physical phenomena, unforeseen in the larger dimensions, such as short-channel effect (SCE) and exponential increase in leakage, are becoming the major bottlenecks for continuous technology scaling. Increasing variations (both inter-die and intradie) in device parameters (channel length, gate width, oxide thickness, device threshold voltage, etc.) produce a large spread in the delay and power consumption in advanced integrated circuits. The presence of large process variations and deep-submicron effects requires a paradigm change in the design and optimisation of large-scale circuits and systems. Innovations only in the area of technology/circuit design are not enough to combat against the different shortcomings of the process variations.

The effects of both within-die and die-to-die process parameter variations will be discussed. The goal of the predictive technology model (PTM) and process compact model (PCM) is to address key design needs, such as variability and reliability, for robust system integration. The predictive technology model, coupled with circuit simulation tools, significantly improves design productivity, providing insight into the relationship between technology/design choices and circuit performance. Use of PTM as a predictive base for exploratory circuit design with extremely scaled CMOS will be shown.

As device physical gate length is reduced, various leakage currents and device parameter variations become the most important considerations for device optimisation. For process engineers, the key tool is the PCM, which provides recipes to meet performance specifications in the face of process variations and dopant fluctuations. TCAD applications include technology and design rule development, extraction of compact models, and more generally, design for manufacturing (DFM). TCAD is expected to be more useful for the 22 nm technology node and beyond, as the capabilities of TCAD can be enhanced to follow the paradigm shifts to processes and materials being considered in nanodevices. In Chapter 9, a methodology for seamless flow of pertinent information between the process and design engineers without the need for disclosing process detail will be presented. Using the global extraction strategy, compact SPICE model parameters have been obtained as a polynomial function of process parameter variations. In this chapter we also discuss the importance of TCAD in constructing process compact models for circuit simulation. A technology-aware circuit design and optimisation technique is discussed. The interactive visual optimisation process using design of experiments (DOE) in a parallel coordinate plot allows one to explore device performance criteria. PCM has been used to find the optimum process conditions to meet a set of device specifications for strain-engineered MOSFETs. Utilisation of TCAD tools for process optimisation for the overall design for manufacturing (DFM) solution is demonstrated.

1.10 Process-Aware Design

The technology choices beyond the 22 nm node will have an ever-increasing impact on the circuit design techniques, and thus an ever closer interaction between the process technologist and circuit designer is required, starting from the technology definition phase. It is now widely recognised that process variation is emerging as a fundamental challenge to integrated circuit design due to CMOS technology scaling, and it will have serious impact on the circuit performance. While some of the negative effects of variability can be handled with improvements in the manufacturing process, the industry has accepted the fact that some of the effects are better mitigated during the design process. Handling device variability during design process will require accurate models of variability and its dependence on designable parameters. In order to continue scaling, there is also a need to reduce margins in the design by classifying process variations as systematic or random.

Chapter 10 provides an overview of current practises as well as near-future research needs in developing a SPICE model for process, variability characterisation, and a process compact model. We present a TCAD methodology of strain-engineered MOSFETs providing the data flow from process simulation to comprehensive and systematic process variability simulation via device simulation, and process compact model analyses to improve design for manufacturing and parametric yield. Basic process-related simulations have been performed using the Sentaurus Process tool to understand the type of stress, device structure, halo implant, gate oxidation, source/drain implant, and annealing temperature. This information is then carried forward to the Sentaurus Device simulation and modelling tool, which provides the electrical characteristics of device structures generated via process simulation. The outputs of the process and device simulations are fed in the process compact modelling framework, i.e., Paramos. Compact SPICE models as a function of process parameter variations are then extracted. The model parameters have been calibrated with the extracted SPICE parameters from TCAD simulations. The usefulness of the process compact model for process optimisation as an overall design for manufacturing solution is shown.

1.11 Summary

Beyond the 22 nm technology node, replacing Si channel with other materials having a higher intrinsic mobility has been proposed. Candidates like Ge and III-V compound semiconductors, e.g., GaAs, are currently being investigated. However, adopting these materials for device fabrication has problems of its own. Special attention should be given to the analysis of performance boosting high-k/metal gate and stress engineering technologies and how they can be used to leverage either speed or leakage or both. New processes in terms of integration of the new substrates based on either a hybrid orientation technique or direct Ge or III-V semiconductors need to be developed. The ultimate scaled MOSFET will be multigate, possibly with advanced and enhanced transport innovations such as III-V channels and nanowire MOSFETs. Finding a compatible gate dielectric, source/drain contact and other integration issues will have to be overcome before they can be used in manufacturing. Compatibility of different mobility enhancement technologies needs to be researched to boost device performance. Challenges for back-end processes include the search for new materials to meet high conductivity and low dielectric permittivity, and interconnect reliability, electromigration, and interaction with assembly and packaging for 3D structures. With the usage of stacked chips and through-silicon vias, stress, thermal, and electrical performance co-analyses will be increasingly needed.

Additional Reading

1. C. K. Maiti, S. Chattopadhyay, and L. K. Bera, *Strained-Si Heterostructure Field Effect Devices*, CRC Press, Boca Raton, FL, 2007.
2. G. A. Armstrong and C. K. Maiti, *Technology Computer Aided Design for Si, SiGe and GaAs Integrated Circuits*, IET, London, UK, 2008.
3. V. Sverdlov, *Strain-Induced Effects in Advanced MOSFETs*, Springer-Verlag, Wien, Austria, 2011.
4. C. Claeys and E. Simoen, *Germanium-Based Technologies: From Materials to Devices*, Elsevier, Amsterdam, 2007.
5. Y. Sun, S. E. Thompson, and T. Nishida, *Strain Effect in Semiconductors Theory and Device Applications*, Springer Science + Business Media, LLC, New York, USA, 2010.
6. C. C. Chiang and J. Kawa, *Design for Manufacturability and Yield for Nano-Scale CMOS*, Springer, Dordrecht, Netherlands, 2007.
7. J.-P. Colinge, Ed., *FinFETs and Other Multi-Gate Transistors*, Springer Science + Business Media, LLC, New York, USA, 2008.
8. Y. Cao, *Predictive Technology Model for Robust Nanoelectronic Design*, Springer Science + Business Media, LLC, New York, USA, 2011.

2

Substrate-Induced Strain Engineering in CMOS Technology

As the conventional metal-oxide-semiconductor field-effect transistor (MOSFET) scaling reached its fundamental limits, several novel techniques have been investigated to extend the CMOS road map. One of these techniques is the introduction of strain in the silicon channel of a MOSFET to obtain higher mobility. Enhancement in mobility may also be obtained by choosing the substrate surface orientation. The major mobility enhancement technologies currently in use can be grouped in two main categories: substrate induced and process induced. Application of strain results in two effects: shift in the band energy and degeneracy splitting of electronic states. Electron mobility is increased by the degeneracy splitting of the conduction band minimum, so the speed of devices fabricated on strained Si is enhanced. One of the predecessors of process-induced strained Si to enhance MOSFET performance is the research that showed enhanced electron mobilities in n-type (100) $Si/Si_{1-x}Ge_x$ multilayer heterostructures and hole mobilities in p-type (100) $Si/i-Si_{1-x}Ge_x/Si$ double heterostructures in early 1980s [1].

The substrate-induced strain techniques use the advantage of either a built-up strain or preferential crystal orientations of the wafer at the process start. Strained Si is one of the key technology boosters identified by the International Technology Roadmap for Semiconductors (ITRS) as being essential to the continuation of classical scaling. The enhanced carrier mobilities made possible through strain engineering result in device performance improvements. However, biaxial stress technology was not adopted in Si CMOS technology due to various issues, which include defects in the substrate, process complexity, cost, and performance loss at high vertical electric fields. Also, strained Si substrates are not yet commercially available with tolerable defect densities. Biaxial strain has also the disadvantage of near-zero hole mobility enhancement at high vertical field, while uniaxial stress shows hole mobility enhancement at large vertical electric fields. Another way of enhancing channel mobility without the introduction of any new channel materials is the use of hybrid crystal orientation of Si substrates. A two to three times boost in hole mobility and about 40–60% improvement in I_{on}/I_{off} performance for p-MOSFET devices is possible by merely changing the starting crystal surface to a (110) Si instead of the typically used (100) Si. Various processing techniques have been proposed to utilise the (100) Si for n-MOSFET and (110) Si for p-MOSFET devices in a silicon-on-insulator (SOI) configuration.

In this chapter, we focus first on the strain-engineered substrates and the principles of strain engineering in Si. Electronic properties of strain-engineered substrates will be covered in Chapter 4. The strain in a Si and SiGe system, a composition of Si with Ge (or C), can provide strain that affects the electrical and optical properties of Si. Strain can be generated from lattice-mismatched film growth in epitaxial heterostructures, intrinsic stress in deposited thin films, and applied external stress. Although many strain technologies have been developed and introduced, they are divided into two distinct categories: global techniques where strain is introduced into the whole wafer and local techniques where stress is delivered to each transistor separately and independently. Local stress is usually introduced during MOSFET fabrication and is also known as process-induced stress, which will be covered in Chapter 3. In this chapter, we also review the major integration challenges and mobility enhancement associated with Ge surface channel devices as well as strained Ge buried channel devices. The smaller band gap (0.67 vs. 1.12 eV for Si) and the much lower melting point (934°C vs. 1400°C for Si) present additional processing challenges for integrating Ge channel MOSFETs. Replacing the channel material is a very significant change from a manufacturing standpoint, and such a modification has not always been successful in volume production for conventional Si CMOS technology. We shall focus on the evaluation of some of the technological alternatives for the integration of channel materials, such as SiGe, Ge, and strained Si in a MOSFET.

2.1 Substrate Engineering

There has been remarkable progress in recent years in Si/SiGe technology. Many novel and advanced devices with high performance have been reported using a SiGe material system [1]. Si and Ge are completely miscible over the entire compositional range and give rise to alloys with a diamond crystal structure. The lattice mismatch between Si and Ge is ~4.17% at room temperature. The lattice constant of $Si_{1-x}Ge_x$ alloys varies linearly, obeying Vegard's rule. When a SiGe alloy layer is deposited on a thick Si substrate, the mismatch is accommodated in either of two ways: tetragonal distortion of the lattice and generation of misfit dislocations at the interface give rise to relaxed or unstrained growth. Initially an epitaxial film of $Si_{1-x}Ge_x$ grown on Si is pseudomorphic; that is, it has the in-plane lattice constant of Si and is compressively strained. However, once the critical thickness for pseudomorphic growth is exceeded, strain is relieved by the formation of dislocations.

One of the most difficult and continuing research challenges in the semiconductor industry is the ability to grow high-quality films using lattice-mismatched materials known as heteroepitaxy. The germanium-silicon system has been extensively studied because of the many potential applications of Ge and advantages over Si. Below a critical thickness, the lattice mismatch between Ge and Si causes the grown film to match the lattice constant of the underlying Si substrate, and hence strain the layer. However, above a critical thickness, it is energetically favourable for the layer to create dislocations to relieve this strain. In addition, due to the lattice mismatch associated with the system, an alternative mechanism of strain relaxation, islanding, often leads to rough surfaces. We describe the main challenges of the SiGe heteroepitaxial system. Different strained and relaxed buffer layers (RBLs) of $Si_{1-x}Ge_x$ (x = 0.13 to 1.00) epitaxially grown on (001) Si substrates are discussed. The main techniques currently used to introduce strain are shown in Figure 2.1. This group includes a wide variety of different wafer types and materials. The wafers can be bulk wafers or SOI based. For p-MOSFETs, mainly compressively strained SiGe layers are used. Ge contents between 20 and 30% and layer thicknesses on the order of 10 nm are required. This provides a stress level in the range 1.5–2 GPa. To achieve tensile strain, virtual substrates (VSs) are normally used. In this case a relaxed SiGe or Ge layer is required on top of a Si wafer. A subsequent deposition of a Si or SiGe layer creates a strained top layer. The strain in the top layer can be tensile or compressive depending

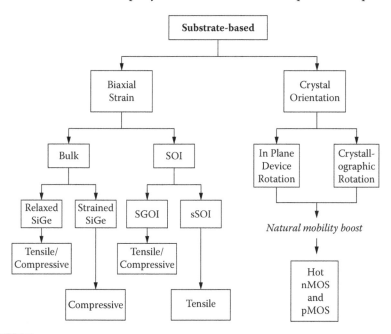

FIGURE 2.1
Different substrate-based mobility enhancement technologies. (After Hallstedt, J., Epitaxy and Characterization of SiGe:C Layers Grown by Reduced Pressure Chemical Vapor Deposition, PhD thesis, Royal Institute of Technology (KTH), Sweden, 2004.)

on the application. For long-channel MOSFET devices, a mobility enhancement up to 10 times for holes and 2 times for electrons has been obtained using this approach [2].

2.2 Strained SiGe Film Growth

The lattice constant of silicon-germanium can be engineered to be close enough to that of silicon so that device quality epitaxial layers can be grown on a silicon (100) surface. The lattice constant of germanium (100) is a_{Ge} = 5.658 Å, and that of silicon (100) is a_{Si} = 5.431 Å. A silicon-germanium alloy, $Si_{1-x}Ge_x$, has a lattice constant, a_{SiGe}, that varies linearly between the lattice constant of silicon and that of germanium. By growing epitaxial SiGe with a considerable amount of silicon, for example, 70% Si, the lattice constant of the $Si_{0.7}Ge_{0.3}$ will be 5.5 Å, which is only 1.2% larger than the silicon lattice constant. A compressively strained pseudomorphic epitaxial layer of $Si_{0.7}Ge_{0.3}$ can be grown on Si (100). As the thickness of the $Si_{0.7}Ge_{0.3}$ layer increases, the amount of strain in the layer increases. Eventually the film reaches a thickness at which its strain energy is greater than the energy needed to form dislocations or change the surface morphology; this thickness is called the critical thickness. Below this thickness the film remains fully strained and does not create defects to relieve that strain.

Many methods have been used for deposition of epitaxial Si and alloys incorporating Ge, C, and Sn on Si substrates. These can be broadly categorised into physical vapour deposition and chemical vapour deposition. Both binary silicon-germanium ($Si_{1-x}Ge_x$) and ternary silicon-germanium-carbon ($Si_{1-x-y}Ge_xC_y$) alloys have found applications in Si CMOS technology. The advances in crystal growth technologies, such as molecular beam epitaxy (MBE), gas source molecular beam epitaxy (GSMBE), organometallic vapour phase epitaxy (OMVPE), and chemical vapour deposition (CVD), have enabled ultra-thin epitaxial semiconductor layers to be routinely grown with both monolayer precision in thickness and composition control to about 1 atomic percent. The main physical vapour deposition method is MBE, which is widely used because of its excellent control over thickness and composition of layers. Two or more of them can form alloys, which have lattice constants between those of the pure form of the constituents. Most of the early work on binary $Si_{1-x}Ge_x$ alloy films was performed using MBE, whereas growth using CVD systems started much later. An epitaxially grown SiGe film on a Si substrate contains a compressive strain because SiGe has a larger lattice constant than Si. Because Ge has an atomic spacing ~4.17% larger than that of Si, the incorporation of Ge into Si could result in a strained SiGe layer. In this case the epitaxial layer of SiGe is in compression in the growth plane to match the substrate lattice atomic spacing. When the first few atomic layers of Ge are deposited, they maintain full bonding with the Si by compressing together. The Si substrate lattice will not be affected because it is much thicker and stiffer. As the Ge content increases the total strain energy increases, and eventually threading dislocations are formed that limit the growth of thicker strained SiGe layers. For given growth conditions, at such a point the strain in the lattice relaxes, and therefore, depending on the Ge content, a critical thickness of SiGe strained layers is set. More about strained layers and theory of critical thickness can be found in reference [2]. Figure 2.2 shows the dependence of the critical thickness on the Ge content for strained $Si_{1-x}Ge_x$ alloys on Si (001). Another factor that causes relaxation of strained layers is the processing temperature. Optimisation of the thermal budget in processing devices

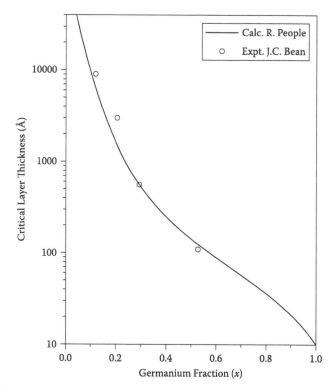

FIGURE 2.2
Dependence of the critical thickness on the Ge content for strained $Si_{1-x}Ge_x$ alloys on Si (001). (After Yousif, M. Y. A., Silicon-Germanium for High-Performance CMOS Technology, PhD thesis, Chalmers University of Technology and Goteborg University, 2001.)

with strained SiGe layers is necessary to avoid relaxation and other deteriorating effects such as Ge segregation. It is important to keep the processing temperature around 600°C or less.

Chemical vapour deposition methods are now available for the growth of very-high-quality strained layers. Notable among them are limited reaction processing CVD (LRPCVD), rapid thermal chemical vapour deposition (RTCVD), and low-temperature ultra-high-vacuum chemical vapour deposition (UHVCVD). Gibbons and his group at Stanford were among the first to demonstrate high-quality $Si_{1-x}Ge_x$ on Si using LRPCVD. The lamp-heated limited reaction processing reactor (LRP) laid the groundwork for other lamp-heated systems at Princeton University and AT&T Bell Laboratories. The UHVCVD reactor pioneered by Meyerson and his coworkers at IBM appeared at nearly the same time as LRPCVD. Combining a standard diffusion furnace with ultra-high vacuum, they have made a very significant impact in growing high-quality alloy layers at low temperature for the fabrication of SiGe heterojunction bipolar transistors (HBTs). A typical UHVCVD system includes a load lock chamber, growth chamber, precursor delivering system, and exhaust of by-product. Chambers are usually pumped by turbo-molecular pump backup by mechanical pump. Inside the process chamber, wafer heating is achieved by carbon susceptor. A typical UHVCVD system provides in situ plasma process capability that can be utilised for plasma-assisted epitaxial growth or low-temperature in situ preclean prior to the epitaxial deposition process. An excellent review of the UHVCVD technique and of the devices fabricated using this method of growth has been published [3].

Because Ge has an atomic spacing ~4.17% larger than that of Si, the epitaxial growth of commensurate $Si_{1-x}Ge_x$ on a relaxed Si substrate, below the critical thickness, would result in a strained $Si_{1-x}Ge_x$ layer. The layer of $Si_{1-x}Ge_x$ will then be under biaxial compressive strain in the growth plane to match the substrate lattice atomic spacing. Figure 2.3 illustrates the strain, misfit dislocation formation, and strain relaxation in $Si_{1-x}Ge_x$ layers grown epitaxially on Si. To achieve electronic grade SiGe heteroepitaxy on Si substrate for channel application, controlling the growth mode is important while achieving target Ge content. Ge content in SiGe film is mainly controlled by controlling the flow rate between Si_2H_6 and GeH_4. For smooth, defect-free 2D film growth, mainly growth temperature and pressure need to be tuned to suppress Stranski-Krastanov (SK) or Volmer-Weber (VW) growth mode, which causes 3D island formation. In a UHVCVD system, reducing growth temperature for high Ge content film in general has been beneficial to enhance 2D growth mode and smoother surface due to reduction of surface diffusivity of adatoms. Epitaxial deposition requires extensive efforts to develop smooth and defect-free (less than $10^6/cm^2$ defect level) to get an optimal mobility and minimal leakage for electronic device application.

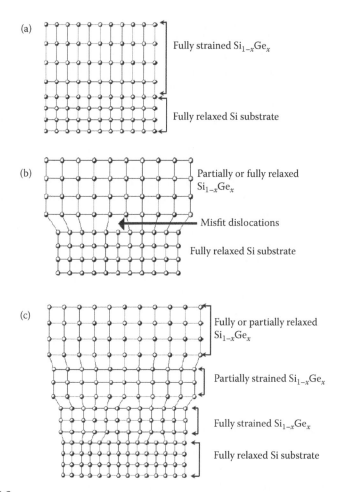

FIGURE 2.3
Illustration of strain (a) misfit dislocation and relaxation above the critical thickness (b) and step-graded technique for virtual substrates (c) in $Si_{1-x}Ge_x$ layers grown epitaxially on Si. (After Yousif, M. Y. A., Silicon-Germanium for High-Performance CMOS Technology, PhD thesis, Chalmers University of Technology and Goteborg University, 2001.)

2.3 Strained SiGe:C Film Growth

Heteroepitaxial SiGe:C layers have attracted serious attention as a material for performance boost in the state-of-the-art MOSFET devices during recent years. Alloying silicon with germanium and carbon adds exclusive opportunities for strain and band gap engineering. The growth of SiGe:C alloys on Si is a challenging task in the sense of finding suitable source materials and growth conditions for a ternary alloy. It is important to investigate the effect of carbon incorporation upon epitaxial growth and its role on strain compensation in $Si_{1-x}Ge_x$ alloys. The major concern is to increase the carbon incorporation rate by avoiding SiC precipitation. To participate in strain compensation, carbon must occupy substitutional sites within the SiGe lattice. Because of the low solubility of carbon within Si, low-temperature growth techniques are required. In the following, some details of SiGe:C epitaxial growth using chemical vapour deposition (CVD) are discussed. Si and Ge can be alloyed over the whole compositional range showing no intermediate phases; on the other hand, C alloying is significantly more complicated. Studies have shown that up to 5% C can be incorporated in $Si_{1-y}C_y$ thin films, which are several orders of magnitude larger than the equilibrium solubility value. Therefore, all layers with C concentrations exceeding the solubility limit are in a metastable state, and special care has to be taken in order to avoid silicon-carbide formation during growth or postannealing treatments.

When a grown layer has a larger or smaller lattice constant than the substrate, then a mismatch system is established with compressive or tensile strain. As a result of the induced strain, an elongation or shrinkage of lattice parameters along the growth direction (out-of-plane) occurs, as shown in Figure 2.4. The relaxation behaviour of these systems is also schematically illustrated. The coordinate system is defined in Figure 2.4, where z denotes the out-of-plane direction and x and y denote the in-plane direction.

2.4 Strained Si Films on Relaxed $Si_{1-x}Ge_x$

The most widely used method to fabricate strained Si is epitaxial growth of Si on a strain-relaxed buffer (SRB) SiGe layer. The relaxed buffer layer (RBL) actually acts as virtual substrates. The strain-relaxed SiGe is created via multilayer compositional grading from pure Si through to the final $Si_{1-x}Ge_x$ alloy composition. This process minimises dislocation nucleation. Substrate-induced strain technologies have focused mainly on biaxial global strain generated by epitaxial growth of a thin Si layer on a relaxed SiGe virtual substrate. The Si layer is biaxial tensile strained in the plane of the interface due to the lattice mismatch between Si and SiGe. A schematic diagram of strained Si heteroepitaxy is shown in Figure 2.5. This result in enhanced carrier transport in the strained Si layer, and mobility enhancements of ~110% for electrons and ~45% for holes have been demonstrated in sub-100 nm strained Si MOSFETs [1].

A tensile-strained Si layer is grown on top of relaxed SiGe. The uniformity of strain in the substrate is crucial to the success because of the high sensitivity of the band structure to strain. The main challenge in fabricating strained Si is the control of defects, in the form of misfit dislocations, which cause variation in strain. The most conventional way to grow relaxed SiGe is to

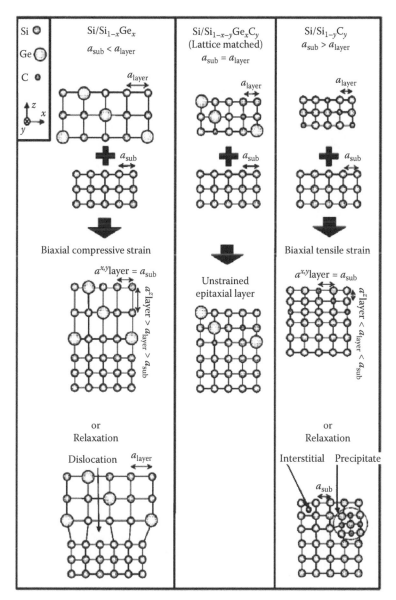

FIGURE 2.4
Schematic presentation of lattice distortion in Si/SiGe:C systems. (After Hallstedt, J., Epitaxy and Characterization of SiGe:C Layers Grown by Reduced Pressure Chemical Vapor Deposition, PhD thesis, Royal Institute of Technology (KTH), Sweden, 2004.)

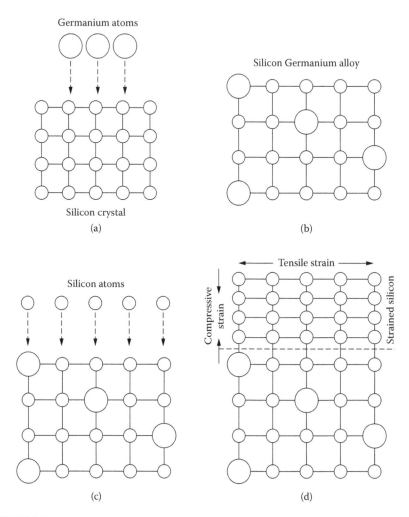

FIGURE 2.5
(a) Introduction of larger Ge atoms into the silicon in order to form alloy; (b) Larger lattice spacing in SiGe alloy in the lateral direction; (c) Introduction of Si on top of the alloy; (d) Formation of a thin layer of strained Si on the SiGe alloy. (After Yousif, M. Y. A., Silicon-Germanium for High-Performance CMOS Technology, PhD thesis, Chalmers University of Technology and Goteborg University, 2001.)

use a graded composition of 1–2 μm/10%Ge. The approach depends largely on materials engineering rather than device design. The strain stretches the silicon lattice by about 1%. This technology is rather mature and dislocation densities on the order of 10^5 cm^{-2} can be obtained. The main drawback of this grading Ge method is the growth of thick buffer layers, which suffers from long deposition times and large material consumption. The epitaxial growth of a thin layer of Si on SiGe alloy is one of the alternatives to strain the silicon, as shown in Figure 2.5. By inserting larger atoms such as Ge into the Si lattice, the spacing between the atoms is increased, as illustrated in Figure 2.5(b). Subsequently, a thin layer of Si atoms is laid on top of this stretched lattice, resulting in the formation of strained Si as depicted in Figure 2.5(d). The thin Si film would thus exhibit biaxial tensile and compressive strain in the lateral and vertical directions, respectively. This approach is attractive for obtaining various degrees of tensile-strained Si. However, high-misfit dislocation density arising from the abrupt transition in lattice spacing exists if the SiGe is grown directly on the Si wafer. These dislocations could penetrate into the surface, and the effect on device electrical performance could be detrimental.

For strained Si, a graded layer of silicon-germanium is grown on top of a bulk silicon wafer. With the step-graded technique, one can obtain a high-quality pure Ge final layer since the underlying buffer of low and intermediate Ge contents acts as a filter to reduce the threading dislocations successively. A typical 2 μm thick SiGe layer having a 20–30% Ge mole fraction, with a higher concentration of Ge atoms at the top, is used. Then a relatively thin layer of silicon, about 20 nm thick, is deposited on top of the $Si_{1-x}Ge_x$ layer. The technology for the growth of a high-quality strained Si layer on a completely relaxed, step-graded, SiGe buffer layer has been reviewed by several authors. Because of the lattice mismatch between Si and $Si_{1-x}Ge_x$, the lattice of the silicon layer is stretched (strained) in the plane of the interface, resulting in enhanced carrier transport in the strained silicon layer, which can be used as the channel of the MOSFET. Strained Si films fabricated by the conventional method using a graded SiGe buffer layer contain dislocations, resulting in nonuniform strain across the wafer. In the direct Ge epitaxy, the ~4.17% lattice mismatch between Si and Ge causes two problems, a high threading dislocation and a high surface roughness due to island formation. The direct epitaxy technique is usually achieved with or without surfactant mediation. In the surfactant-mediated epitaxy, a monolayer of a surfactant is first deposited. This surfactant saturates the dangling bonds of the semiconductor surface, and consequently reduces the surface free energy for both Ge and Si. For details, the reader may refer to the Special Issue on Strained-Si Heterostructures and Devices [4].

2.5 Strained Si on SOI

The use of SOI substrates is another method to create virtual substrates. This is based on oxidation of a sacrificial SiGe layer grown on a Si body. In this case, the oxidation of SiGe layers (usually with layer thickness of 20–50 nm) is at high temperature (1050–1150°C). Then the oxidation process favours the Si atoms and the Ge atoms are diffused down. The buried oxide (BOX) layer of SOI acts as a diffusion barrier leading to a condensation of Ge at the interface. As a result, a thin relaxed SiGe layer is obtained that is suitable for deposition of a tensile Si layer. This layer can also be transferred to a new oxidised Si wafer, which leads to a so-called strained SOI (SSOI) wafer [1].

From the viewpoint of the scaling, it is preferable that the strained Si layer is directly formed on SiO_2/Si substrates without utilising SiGe buffer layers. Strained silicon-on-insulator (SSOI) is a material system that combines the carrier transport advantages of strained Si with the reduced parasitic capacitance and improved MOSFET scalability of thin-film SOI. Also, SOI technology provides many significant improvements (over bulk Si CMOS), for example, in minimising parasites, decreasing leakage, improving short-channel effects (SCEs), facilitating better noise isolation, and improving single-event upset (SEU) tolerance. The performance benefit of combining strained silicon with an SOI has been demonstrated in a 60 nm gate length, n-channel MOSFET with ultra-thin thermally mixed strained silicon/SiGe-on-insulator substrate, such as strained silicon-on-insulator (SSOI) and SiGe-on-insulator (SGOI). Twenty to 25% drain current enhancement has been demonstrated at short-channel length [5]. Recently, transistors using ultra-thin strained silicon-directly-on-insulator (SSDOI) have been demonstrated (see Figure 2.6) that eliminate the SiGe layer before transistor fabrication, thereby providing higher mobility, while eliminating the SiGe-induced material and process integration problems.

An SSDOI structure is fabricated by a layer transfer or wafer bonding technique. First, an ultra-thin layer of strained silicon is formed epitaxially on a relaxed silicon-germanium layer, and an oxide layer is formed on top. After hydrogen is implanted into the SiGe layer, the wafer is flipped and bonded to a handle substrate. A high-temperature process splits away most of the original wafer and leaves the strained silicon and SiGe layers on top of the oxide layer. The SiGe is then selectively removed, and transistors are fabricated on the remaining ultra-thin strained silicon. Both electron and hole mobility enhancement have been observed in an SSDOI structure indicating that strain is retained after the complete device processing steps. Using

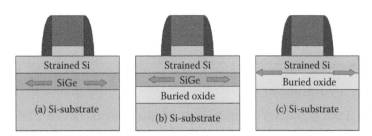

FIGURE 2.6
(a) Strained Si/SiGe on bulk wafer. (b) SiGe-on-insulator (SGOI) MOSFET. (c) Strained Si-directly-on-insulator (SSDOI). (After Maiti, T. K., Process-Induced Stress Engineering in Silicon CMOS Technology, PhD thesis, Jadavpur University, 2009.)

SmartCut and wafer bonding techniques, global strain can also be integrated on SOI substrates.

The major drawback of all global strain techniques for CMOS technology is that they can provide only one type of strain. Since the mobilities of electrons and holes are differently affected by different strains (compressive or tensile), a global strain configuration, for example, compressive biaxial strain, can be beneficial for p-MOSFETs, but deteriorate the n-MOSFET performance. This problem is circumvented by local strain techniques, which are able to provide different strain patterns in n- and p-MOSFETs. The development of alternative growth techniques for the deposition of good quality Ge and strained Si without the growth of thick relaxed SiGe buffer layers is very important from a cost standpoint, and such techniques would hopefully make these materials suitable for volume production.

2.6 Strained Ge Film Growth

Germanium offers higher mobility for both electrons (factor of 2) and holes (factor of 4) than silicon. MOSFETs on Ge bulk wafer were hindered for decades because of the lack of a stable native oxide, in contrast to its counterpart silicon. While previous Ge channel transistors were predominantly on Ge bulk wafers, integration into Si is preferred for CMOS compatibility. Compressive Ge provides a large enhancement to hole mobility. Ge channel structure originally has higher hole mobility than strained Si, and the larger hole mobility of germanium mainly comes from the smaller effective mass of the holes. Germanium-on-insulator (GeOI) MOSFETs have become promising for monolithic 3D ICs owing to their low processing temperatures. Ge channels are grown on $Si_{1-x}Ge_x$ buffer layers. Some promising results have been obtained by the use of a dislocation blocking layer stack composed of multiple SiGe layers. Compressive strain introduced into the Ge channel can enhance the mobility beyond bulk Ge, since the hole effective mass is reduced and interband phonon scattering is suppressed due to the splitting of light-hole (LH) and heavy-hole (HH) bands. In silicon MOSFETs, biaxial tensile strain is obtained via applying $Si_{1-x}Ge_x$ substrate underneath the Si channel. Biaxial tension is not a popular stress type for germanium devices due to the large lattice constant of germanium. Biaxial compressive stress in the germanium MOSFET channel can be obtained by the germanium channel on top of the $Si_{1-x}Ge_x$ substrate.

Unstrained Ge hole mobility vs. vertical electric field and device surface orientation is shown in Figure 2.7. Strain-enhanced hole mobility in silicon and germanium p-MOSFETs has been reported [6]. k.p calculations are commonly used to give physical insights into hole mobility enhancement at large stress (3 GPa for Si and 6 GPa for Ge) for stresses of technological

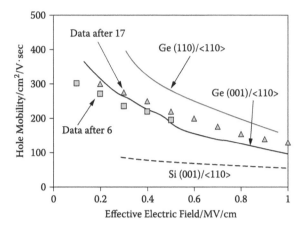

FIGURE 2.7
Germanium hole mobility vs. effective electric field. (After Sun, G., Strain Effects on Hole Mobility of Silicon and Germanium p-Type Metal-Oxide-Semiconductor Field-Effect-Transistors, PhD thesis, University of Florida, 2007.)

importance: in-plane biaxial and channel direction uniaxial stress on (001) and (110) surface-oriented p-MOSFETs with <110> and <111> channels. The hole mobility vs. uniaxial compressive stress for Ge is shown in Figure 2.8 for (001)-oriented Ge. For (001)-oriented devices, both Si and Ge show large enhancement. One difference between the two curves is that the mobility enhancement for Si saturates at about 3 GPa, but it does not saturate until 6 GPa of stress is applied to Ge.

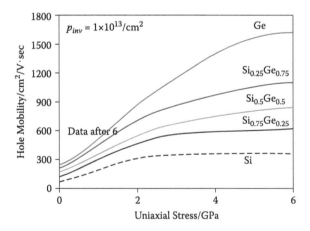

FIGURE 2.8
Germanium and silicon hole mobility on (001)-oriented device under uniaxial compressive stress where the inversion hole concentration is 1 ′ 10¹³/cm². (After Sun, G., Strain Effects on Hole Mobility of Silicon and Germanium p-Type Metal-Oxide-Semiconductor Field-Effect-Transistors, PhD thesis, University of Florida, 2007.)

Growth techniques commonly used for compressively strained Ge films include both reduced pressure chemical vapour deposition (RPCVD) and ultra-high-vacuum chemical vapour deposition (UHVCVD). As the first step, low-temperature RPCVD is used to grow a fully relaxed SiGe virtual substrate layer at 500°C with a thickness of ~135 nm, surface roughness of 0.3 nm, and Ge content of 77%. Then, low-temperature UHVCVD was used to grow a high-quality strained pure Ge film on the SiGe virtual substrate at 300°C. Finally, a very thin strained Si layer of 1.5–2 nm thickness was grown on the Ge layer at 550°C for the purpose of passivation and protection. The whole epitaxial layer thickness is less than 150 nm. Due to the low growth temperature, the two-dimensional layer-by-layer growth mode dominates during the epitaxial process, which is a key factor for the growth of high-quality strained Ge films.

2.7 Strained Ge MOSFETs

Although Ge has a low effective mass for electrons providing for higher injection velocities, it also has a high dielectric constant and smaller band gap, making it susceptible to higher leakage and worse short-channel effects. One major problem for Ge CMOS device fabrication is that it is very difficult to obtain a stable oxide gate dielectric. Poor chemical and mechanical stability prohibits the use of germanium dioxide (GeO_2) as a gate dielectric for Ge devices. The water-soluble native Ge oxide that is typically present on the upper surface of a Ge-containing material causes this gate dielectric instability. The best-known dielectric candidate for use on Ge is Ge oxynitride (GeO_xN_y). High-quality thin GeO_xN_y can be formed on germanium by nitridation of a thermally grown germanium oxide. Rapid thermal oxidation (RTO) at 500–600°C followed by rapid thermal nitridation (RTN) at 600–650°C in ammonia (NH_3) ambient has generally been practised. Also, high-quality thin GeO_xN_y could serve as a stable interlayer for integration of novel high-k dielectrics into Ge MOS devices. Recent studies on high-k dielectrics for silicon MOSFETs by ALD and MOCVD techniques have prompted activities to develop Ge MOSFETs implementing high-k dielectrics such as ZrO_2 and HfO_2 (binary metal oxides). A Ge channel transistor integrated on a Si wafer suffers from high-density defects in the Ge epilayer and poor surface roughness due to ~4.17% lattice mismatch between Si and Ge. However, one of most challenging tasks for Ge/high-k MOS systems is the Ge surface preparation and interface control before high-k film deposition. It appears essential to have a surface free of germanium oxide before high-k film deposition. With the development of high-k dielectric, Ge MOSFETs with high-k/metal gate using different passivation methods have been successfully demonstrated on bulk Ge wafers. Because of the low melting point of Ge, it is desirable to use metal gate electrodes rather than conventional poly-Si gate electrodes where high-temperature (>900°C) dopant activation is required. Metals such as Al, W, Pt, TiN, and TaN are among the most popular metal electrodes reported for Ge MOSFETs. The metal gate electrodes are chosen considering their interaction with the Ge gate dielectric. The low band gap of germanium (0.67 eV compared with 1.12 eV for Si) presents a device design challenge.

Surface channel Ge MOSFETs have been demonstrated using thin Ge oxynitride or high-k dielectric as gate insulator. Strained Ge p- and n-MOSFETs on relaxed $Si_{1-x}Ge_x$ graded buffers have been reported [7]. To accommodate the wafer incompatibility, Ge-based devices in this research are fabricated on relaxed SiGe grown on Si wafers, and to avoid the use of GeO_2, a thin epitaxial Si layer is grown on top of the strained Ge channel. The Si cap allows a high-quality interface to be formed with a conventional SiO_2 gate and ensures basic compatibility with conventional Si CMOS processing. For strained Ge layers on relaxed SiGe, the valence band is offset from the relaxed virtual substrate below the channel and the Si above, resulting in a well for holes. Furthermore, compressive strain reduces the hole effective mass and lifts the valence band degeneracy in Ge. The smaller band gap in Ge has been a concern because of its influence on junction leakage and band-to-band tunneling. The junction leakage of n^+/p and p^+/n Ge diodes formed by boron and phosphorus implantation can be reduced to ~10^{-4} A/cm^2 with annealing. This is considered acceptable for device operation. It has been shown that the band-to-band tunneling can be reduced dramatically through careful device structure design.

Monolithic integration of tensile-strained Si/germanium (Ge) channel n-MOSFET and tensile-strained Ge p-MOSFET with ultra-thin (equivalent oxide thickness ~ 14 Å) HfO_2 gate dielectric and TaN gate stack on Si substrate has been demonstrated [8]. Defect-free Ge layer (279 nm) grown by ultra-high-vacuum chemical vapour deposition is achieved using a two-step Ge growth technique coupled with compliant Si/SiGe buffer layers. The epi-Ge layer experiences tensile strain of up to ~0.67% and exhibits a peak hole mobility of 250 cm^2/V·s, which is 100% higher than the universal Si hole mobility. The gate leakage current is two orders of magnitude lower than the reported results on Ge bulk. A modified two-step growth of a Ge layer using an intermediate ultra-thin SiGe buffer and compliant Si epilayer occurs in an ultra-high-vacuum (UHV) chemical vapour deposition chamber. A 279 nm thick Ge layer with a very low threading dislocation of 6×10^6 cm^{-2} is realised on a Si substrate. We also demonstrate n- and p-MOSFETs on tensile-strained Si and Ge (s-Si/s-Ge) with HfO_2/TaN gate stack on the Si substrate. The cross-sectional high-resolution transmission electron microscopy photograph of the Si/SiGe/Ge heterostructure in Figure 2.9(a) shows that the dislocations are mainly confined within the SiGe buffer layer and at the SiGe/Ge interface. The threading dislocation density of the Ge layer above the Ge/SiGe interface is less than 10^7/cm^2. Surface roughness measurement using atomic force microscopy on epi-Ge shows a root mean square value of 0.425 nm for 10×10 μm^2 pad. The epi-Ge thickness is about 279 nm; therefore,

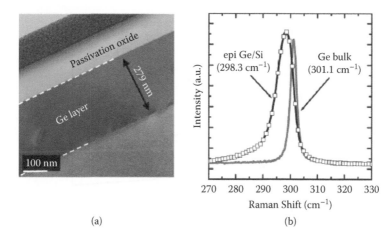

(a)   (b)

FIGURE 2.9
(a) HRTEM image of the heterostructure epitaxial layers of Si/SiGe/Ge layers. Misfit dislocations are confined within the SiGe buffer and at the interface of SiGe/Ge. (b) Raman spectra
for epi-Ge on Si wafer and Ge bulk. Phonon peak for Ge on Si downshifts, indicating tensile
strain in Ge. Spectra broadening for Ge on Si is due to Ge intermixing with Si cap. (After Zang,
H., W. Y. Loh, J. D. Ye, G. Q. Lo, and B. J. Cho, *IEEE Electron Dev. Lett.*, 28, 1117–1119, 2007. With
permission.)

the fabricated MOSFETs are far from a defect-rich region. Due to epi-Si pas-
sivation on Ge, the leakage current of HfO_2 is 10^{-4} to 10^{-5} A/cm^2 at $|V_g| = 1$ V.
Figure 2.10 shows the DC characteristics of the Ge CMOSFETs fabricated on Si/
SiGe/Ge/s-Si substrate. The extracted threshold voltages are 0.25 and –0.35 V
for n- and p-MOSFETs, respectively. The $I_d - V_d$ curves of Ge CMOS with a gate
length of 5 μm show an excellent performance in p-MOSFET, as expected. Drain

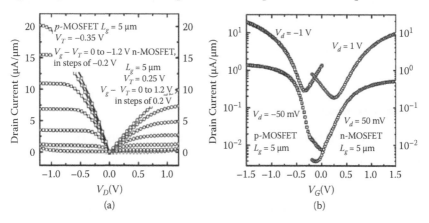

(a)   (b)

FIGURE 2.10
(a) I_d-V_d and (b) I_d-V_g characteristics for p- and n-MOSFETs on strained Si/tensile-strained Ge.
(After Zang, H., W. Y. Loh, J. D. Ye, G. Q. Lo, and B. J. Cho, *IEEE Electron Dev. Lett.*, 28, 1117–1119,
2007. With permission.)

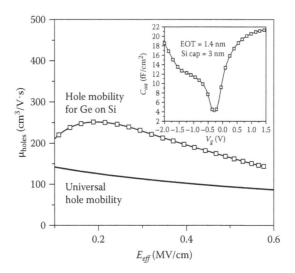

FIGURE 2.11
Extracted hole mobility for Ge p-MOSFET measured using split C-V method. Twice the hole mobility improvement is achieved compared to Si universal mobility. (After Zang, H., W. Y. Loh, J. D. Ye, G. Q. Lo, and B. J. Cho, *IEEE Electron Dev. Lett.*, 28, 1117–1119, 2007. With permission.)

currents are $20\,\mu A/\mu m$ for p-MOSFET and $10\,\mu A/\mu m$ for n-MOSFET at $|V_g - V_T|$ = 1.2 V. n-Channel MOSFETs show a lower drive current than p-MOSFETs, and this may be attributed to intervalley scattering and defect scattering in the Si cap layer. Extracted hole mobility for Ge p-MOSFETs measured using the split Capacitance-Voltage (C-V) method is shown in Figure 2.11. The C-V characteristic measured on Ge p-MOSFET is shown in the inset. Since germanium is well known to have the highest hole mobility among all semiconductors, the p-MOSFET would have a big chance for improvements. Additionally, the SiGe/Si material system improves not only the hole mobility but also the electron mobility, i.e., better CMOS performance.

2.8 Heterostructure SiGe/SiGe:C Channel MOSFETs

The performance of conventional CMOS circuits is primarily limited by the lower transconductance of the p-MOSFET, compared to the n-MOSFET, because the field-effect hole mobility is about three times lower than that of the electron. To minimise this asymmetry and to improve the current drivability, the p-MOSFET needs to be designed with a large size compared to the n-MOSFET, thus affecting the packing density and speed. Silicon-germanium (SiGe) strained layers have shown promising results for device applications. The driving forces have been to make new devices, and the key to success has been the possibility of band gap engineering in silicon-based materials. The SiGe technology is expected to boost the performance of Si-based devices beyond that of Si. The gain in performance can be obtained even with less aggressive downscaling, and thus improved reliability. In the following, we discuss the channel-engineered MOSFETs using different structures of $Si/Si_{1-x}Ge_x/Si$ quantum wells and explain how a strained $Si_{1-x}Ge_x$ layer of a few nanometers thick (typically 5–10 nm) epitaxially grown on a Si substrate can be used as a quantum well to confine holes. The application of strained layers to heterostructure FETs is not as well developed as HBTs. Although research in $Si_{1-x}Ge_x$ channel heterostructure MOSFETs is gaining momentum, they are not expected to be in the Si-CMOS mainstream before conventional Si counterparts reach the fundamental limit.

Though there is less development on SiGe heterostructure field-effect transistors (HFETs) than on HBTs, they are very attractive because they are compatible with standard Si CMOS technology and can perform better and more reliably even with less downscaling. Despite the advantages of speed, low noise, and low-power-delay product, the $Si_{1-x}Ge_x$ CMOS technology is not free from drawbacks. These drawbacks include a high leakage current from the source/drain (S/D) to the substrate in devices fabricated on relaxed $Si_{1-x}Ge_x$ virtual substrates due to the narrower band gap of the substrate. The film growth for $Si/Si_{1-x}Ge_x/Si$ strained layers has been discussed earlier. Strained SiGe layers on Si substrates can be used as quantum well (QW) channels to confine holes in p-MOSFETs with enhanced hole mobility compared to conventional Si MOSFETs. Different structures of $Si/Si_{1-x}Ge_x/Si$, for example, single (SQW) and double (DQW) quantum well p-MOSFETs, are considered. Threshold voltage, charge control, and short-channel effects have been studied for these structures. A strained SiGe layer may be used as a quantum well to engineer the channel of the p-MOSFET, and SiGe polycrystalline thin film may be used to engineer the gate of the transistor.

2.8.1 Band Alignment

Band engineering of Si and Ge can be used to produce band discontinuities in the valence and conduction bands. Therefore, electrons and holes can be confined in quantum wells. The strain plays a dominant role in determining the alignment of bands at heterointerface, thus determining the confinement energy of hole/electrons in the quantum wells. When a $Si_{1-x}Ge_x$ layer is grown pseudomorphically on a relaxed Si substrate, a biaxial compressive strain will take place in the $Si_{1-x}Ge_x$ layer, which then serves as a quantum well for holes. On the other hand, if a Si layer is grown epitaxially on a relaxed $Si_{1-x}Ge_x$ layer, a biaxial tensile-strained Si layer will be formed, which can then act as a quantum well for electrons (see Figure 2.12). For the biaxial compressive strain, the in-plane lattice parameters become smaller to be in registry with the underlying substrate, and the perpendicular lattice parameter becomes larger. For the biaxial tensile strain, the in-plane lattice

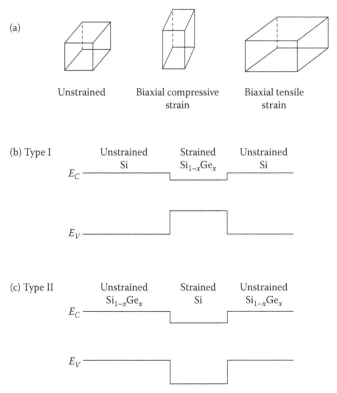

FIGURE 2.12
Illustration of biaxial compressive and tensile strains (a), and energy band diagrams of (b) a strained $Si_{1-x}Ge_x$ layer sandwiched between unstrained Si layers (compressive) and (c) a strained Si layer sandwiched between unstrained $Si_{1-x}Ge_x$ layers (tensile). (After Yousif, M. Y. A., Silicon-Germanium for High-Performance CMOS Technology, PhD thesis, Chalmers University of Technology and Goteborg University, 2001.)

parameters stretch out to match the substrate, while the perpendicular lattice parameter becomes smaller. In the former case the band alignment is of type I, and the band offset lies predominantly within the valence band, and in the latter one the band alignment is of type II.

A composition of Si and Ge will have a band gap between the elemental values. SiGe alloys have an asymmetrical band gap where the main change in the band gap (band offset) is in the valence band side. The induced compressive strain splits the LH and HH together with spin-orbit bands. The HH band is shifted upwards and changes the curvature into a light electron-like band. The original LH band also changes its character to become a heavy-hole-like band. The induced strain shifts furthermore downward the spin-orbit splitting band. These changes of the band gap indicate that the SiGe/Si system can be used to enable confinement of a hole inversion layer in the SiGe of a p-MOSFET device. Enhancement of both hole and electron has been reported by having biaxial tensile Si as a channel layer in the transistors. The main reason for higher mobility is reduction of intervalley and interband phonon scatterings. Although biaxial tensile strain increases the hole mobility, this effect is diminished at high vertical electric fields.

In the biaxially strained SiC/Si system, the amount of strain is increased by increasing the C content. The exact effect of carbon in fully or partially strain-compensated SiGe:C layers is not well known. Mainly, the band splitting is believed to decrease as the strain is compensated. However, the incorporation of small amounts of C into the $Si_{1-x-y}Ge_xC_y$ matrix leads to an increase of the band gap. In Figure 2.13, the effect of the strain on the fundamental band gap is displayed. The hashed region reflects uncertainties in measured values of the deformation potential. It is evident from this figure that as the Ge content is increased, the band gap of the strained alloy decreases.

2.8.2 Mobility Enhancement

The mobility enhancement in strained $Si_{1-x}Ge_x$ layers is mainly, due to the presence of Ge and the confinement in the valence band. The strain causes a

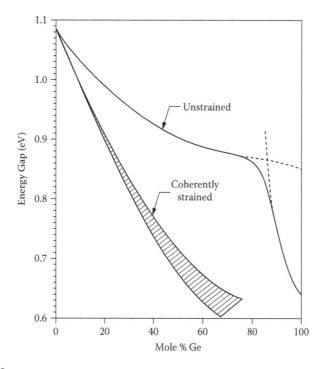

FIGURE 2.13
Fundamental (lowest-energy) indirect band gap of strained and unstrained alloys on Si(001) substrates. (After Yousif, M. Y. A., Silicon-Germanium for High-Performance CMOS Technology, PhD thesis, Chalmers University of Technology and Goteborg University, 2001.)

reduction in the traverse effective mass in strained $Si_{1-x}Ge_x$ layers. The compressive strain in the plane also lifts the degeneracy of the valence band, and the HH band dispersion becomes LH-like, and therefore interband scattering is reduced. Increasing the Ge percent also results in further reduction of the effective mass. For lateral transport, it is desirable to have a light in-plane effective mass in order to enhance the carrier mobility. In a p-MOSFET with such structures, an intrinsic buried $Si_{1-x}Ge_x$ channel is desirable to reduce the Coulomb scattering of carriers by ionised impurities. On top of this layer a thin cap layer of Si is deposited to facilitate growing the gate oxide (see Figure 2.14). Thermal oxidation of strained $Si_{1-x}Ge_x$ is not viable and results in a pile-up of Ge at the dielectric/SiGe interface with an increase of interface state density, thus degrading the mobility. It should be noted that one of the advantages of buried channels is that they provide more immunity to hot-carrier degradation because carriers have to travel longer distances to reach the gate dielectric. The transconductance reduction in buried channels is already compensated for by the higher hole mobility in strained SiGe materials, as we explained earlier, and also by avoiding the Si/SiO_2 interface roughness scattering.

2.8.3 Double Quantum Well p-MOSFETs

In a Si/SiGe channel p-MOSFET, a strained $Si_{1-x}Ge_x$ layer is epitaxially grown a few nanometers (~2 nm) below the gate oxide (see Figure 2.14). This layer is grown directly on a Si substrate and serves as a quantum well to confine holes, because essentially all of the band gap difference is incorporated in the valence band. The enhancement of hole mobility in strained SiGe layers is, in part, due to the reduction in the hole effective mass in the transverse direction. In addition, the compressive strain in the plane lifts the valence band degeneracy, and the HH band dispersion becomes LH-like. Although improvement has been achieved in SQW SiGe p-MOSFETs, at high gate bias the Si surface layer beneath the gate oxide will also be populated by holes, and eventually the SQW p-MOSFET will operate as a conventional Si MOSFET.

As far as short-channel effect is concerned, SiGe MOSFETs also show better performance over conventional Si devices in the deep submicron regime. It has been shown that the velocity in 0.1 µm Si/SiGe MOSFETs is higher, and that velocity overshoot occurs closer to the source end of the device, compared to conventional MOSFETs, to have the higher performance. Therefore, with such promises of performance in the deep submicron regime, SiGe devices may be regarded as an alternative to scaling. Buried channel strained $Si_{1-x}Ge_x$ SQW p-MOSFETs of enhanced performance, compared to control Si devices, have been reported. Hot-carrier degradation in device characteristics of conventional Si p-MOSFETs is caused mainly by trapped electrons in the gate oxide or hot-carrier-induced interface traps between the gate oxide and the surface silicon. For a SiGe device, heterointerface traps are induced by hot carriers. Comprehensive reviews of strained layer quantum wells (QWs) of $Si_{1-x}Ge_x$ and Si can be found in [1].

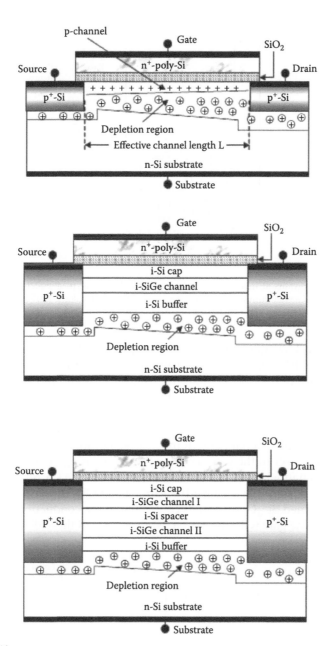

FIGURE 2.14
Schematic diagram illustrating the (a) Si p-MOSFET, (b) SQW SiGe p-MOSFET, and (c) DQW SiGe p-MOSFET. (After Yousif, M. Y. A., Silicon-Germanium for High-Performance CMOS Technology, PhD thesis, Chalmers University of Technology and Goteborg University, 2001.)

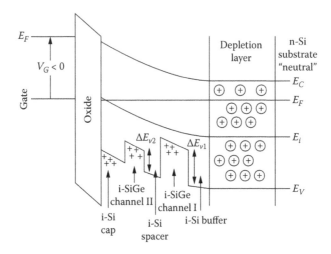

FIGURE 2.15
Energy band diagrams at inversion for retrograde DQW $Si_{1-x}Ge_x$ p-MOSFETs. (After Yousif,
M. Y. A., Silicon-Germanium for High-Performance CMOS Technology, PhD thesis, Chalmers
University of Technology and Goteborg University, 2001.)

In their work on the DQW SiGe, Yousif [9] extended the possibilities of
$Si_{1-x}Ge_x$ MOSFETs for further improvements. In Figure 2.14(c), the schematic diagram of the double quantum well (DQW), and in Figure 2.15, the
energy band diagram of this device at inversion is shown. Different designs
with different Ge contents were reported, and the Ge profiles for the different structures in DQW $Si/Si_{1-x}Ge_x$ MOSFET are shown in Figure 2.16. In
design I, the high-Ge channel will invert first, because it has a larger band
offset and is closer to the gate. Therefore, the screening effect may prevent
the low-Ge channel from reaching strong inversion. As a result, a device
with such a design may behave the same as a SQW device. In design II,
both the strained $Si_{1-x}Ge_x$ channels contribute to the conduction. The channel closer to the gate will invert first. Note that in all these designs, shown

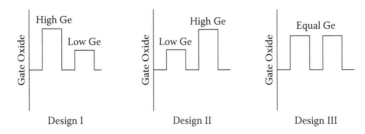

FIGURE 2.16
The Ge profiles for the different structures of DQW $Si/Si_{1-x}Ge_x$ MOSFET. (After Yousif,
M. Y. A., Silicon-Germanium for High-Performance CMOS Technology, PhD thesis, Chalmers
University of Technology and Goteborg University, 2001.)

in Figure 2.16, a thin Si spacer separating the two QWs is desirable as far as mobility is concerned, but also, the QWs should be maintained uncoupled, because such communication, i.e., the overlap of the carrier wave function, will degrade the mobility. In the retrograde DQW of design II, the high-Ge channel reaches strong inversion first, followed by the low-Ge channel, and at higher gate biases the surface channel also inverts. The main conducting channel is the high-Ge channel because it populates higher hole density. Since this main channel is more buried than in SQW, it should come as no surprise that hot-carrier degradation is further suppressed. It is expected that the retrograde DQW structure will have a further improvement in mobility, hot-carrier degradation, transconductance at low-voltage operation, and a reduced $1/f$ noise and random telegraph signals. Such a structure may be suitable for analogue applications. Recently $Si_{1-x}Ge_x$ p-MOSFETS with a retrograde double quantum well structure have been demonstrated experimentally with improved performance.

The turn-on of the surface channel can be suppressed by thinning the Si cap layer. In the case of SQW devices, the SQW channel comes closer to the gate and an improved transconductance can be expected. However, the expected gain in transconductance is not achieved in these devices due mainly to surface roughness and interface scattering. Also, SQW MOSFETs are susceptible to both hot-carrier degradation and mobility degradation. This trade-off might be less problematic in the retrograde DQW device because the main high-Ge channel is separated from the gate by the other low-Ge channel. As for hole density, it has been reported that there is no improvement in the retrograde DQW structure, compared to design I, which almost acts as a SQW device.

The intrinsic material properties of $Si_{1-x}Ge_x$, rather than shrinkage of devices, make possible the achievement of performance enhancement in these devices. Employment of <100> channel direction in a strained $Si_{0.8}Ge_{0.2}$ p-MOSFET has demonstrated the substantial amount of hole mobility enhancement, as large as 25%, and parasitic resistance reduction of 20%, compared to a <110> strained $Si_{0.8}Ge_{0.2}$ channel p-MOSFET, which already has an advantage in mobility and the threshold voltage roll-off characteristic over the Si p-MOSFET. This result indicates that the <100> strained SiGe channel p-MOSFET is a promising and practical candidate for realising high-speed CMOS devices under low-voltage operation. In general, in comparison with conventional Si MOSFETs, $Si/Si_{1-x}Ge_x/Si$ MOSFETs benefit from the following advantages: (1) higher channel mobility, (2) smaller width, i.e., higher packing density, (3) higher transconductance and improved speed, (4) lower-power-delay product, (5) better immunity to hot-carrier degradation, and (6) reduced flicker and random telegraph noise.

SiGe technology, developed for over two decades, has been plagued by the problem for device applications requiring a high Ge mole fraction. The thermal stability of strained Si and compressively strained $Si_{1-x}Ge_x$ layers is a major concern in many device structures. Consequently, the design flexibility is limited for applications involving a low Ge concentration (in the buffer layer in the case of strained Si), a thinner active layer, and relatively low process temperature windows. Incorporating smaller-size carbon atoms substitutionally into the SiGe system enables one to compensate the strain, which leads to an increase in thermal stability and critical layer thickness. Growth of strain-compensated ternary SiGe:C layers and relaxed buffer layers (as a template for growing strained Si) has been reported by many researchers [2].

Incorporation of carbon has paved the way to extending SiGe-based heterostructures, allowing more flexibility in strain and band gap engineering. Carbon-containing alloys promise to expand the range of device applications of silicon-based heterostructures. It has been shown that carbon reduces the strain in SiGe at a faster rate than it increases the band gap. Thus for a given band gap, a larger critical thickness can be obtained for $Si_{1-x}Ge_xC_y$ films than for those without carbon. The extracted valence band offset in $Si_{1-x}Ge_xC_y$ heterostructures also decreases much more slowly than predicted with increasing carbon in the alloy, so that for a given lattice mismatch to Si, the valence band offset is larger for SiGe:C than for Si. Therefore, both the band alignment and the valence band offset in ternary alloys are favourable for various device applications, as they reduce the possibility of process-induced strain relaxation, while confining the holes in the valence band quantum well. Partially strain compensated $Si_{0.793}Ge_{0.2}C_{0.007}$ p-MOSFET devices have been fabricated using UHVCVD grown layers [10]. The devices show good linear and saturation characteristics. Enhanced performance of ternary devices at room temperature has been reported. The ternary device, however, shows a lower mobility at 77 K than the binary device due to increased alloy and surface roughness scattering. The alloy scattering potential and field-dependent mobility degradation factor of the ternary SiGe:C layer have been estimated.

2.9 Strained Si MOSFETs

While biaxially compressed $Si_{1-x}Ge_x$ offers many desirable properties, most of the advantages are encountered in the valence band causing an enhancement in hole mobility. To realise improvements in electron mobility and a usable conduction band offset, it is necessary for the material to be in biaxial tension. As discussed earlier, a smaller lattice constant Si epilayer will be in biaxial tension when grown on a relaxed $Si_{1-x}Ge_x$ with larger lattice constant. In this case, type II band offset occurs and the structure has several advantages over the more common type I band alignment, as a large band offset (on the order of 100 meV or more) is obtained in both the conduction and valence bands, relative to the relaxed $Si_{1-x}Ge_x$ layer. Strained Si provides

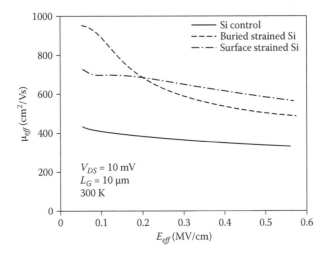

FIGURE 2.17
Comparison of effective electron mobilities between the surface channel, buried channel strained Si n-MOSFETs, and Si control n-MOSFETs. (After Welser, J. J., The Application of Strained-Silicon/Relaxed-Silicon Germanium Heterostructures to Metal-Oxide-Semiconductor Field-Effect Transistors, PhD thesis, Stanford University, 1995.)

both larger conduction and valence band offsets and does not suffer from alloy scattering (hence mobility degradation). The significant improvement in both electron and hole mobility shows the possibility of both n- and p-type FET devices for strained Si/SiGe-based heterostructure CMOS (HCMOS) technology.

First-generation strained Si MOSFETs were all based on SiGe virtual substrates. The strained Si n-MOSFET grown on SiGe virtual substrate to improve electron mobility was first demonstrated by J. J. Welser in 1992. In the experiment, strained Si channel was grown on relaxed $Si_{0.71}Ge_{0.29}$ layers, which were on top of the graded buffer layer. The mobility of the strained and unstrained n-MOSFETs is shown in Figure 2.17, where the peak mobility at room temperature was enhanced by about 2.2 times. For the surface channel strained Si device mobility is enhanced compared to the control Si device and has a similar dependence on the effective electric field. The peak mobility is 1,000 cm^2/Vs, which shows an 80% enhancement over control Si devices (550 cm^2/Vs). The peak mobility value for buried channel devices is over 1,600 cm^2/Vs, which is almost three times that of the control Si device. Room temperature effective mobility vs. electric field curves of surface channel strained Si n-MOSFETs with different Ge contents in the buffer layer is shown in Figure 2.18 along with the mobility extracted from a control Si device. Strained Si mobility increases with increasing strain (more Ge content in the relaxed buffer layer) and has little dependence on the effective electric field.

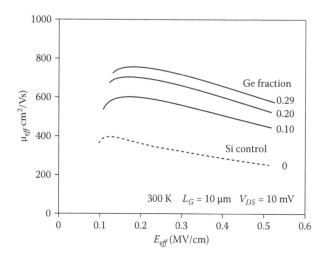

FIGURE 2.18
Effective mobility of surface channel strained Si n-MOSFETs at room temperature. All curves are parallel and the mobility increases with increasing strain in the Si channel as measured by the increasing apparent Ge fraction in the relaxed buffer layer indicated. (After Welser, J. J., The Application of Strained-Silicon/Relaxed-Silicon Germanium Heterostructures to Metal-Oxide-Semiconductor Field-Effect Transistors, PhD thesis, Stanford University, 1995.)

Observation of hole mobility enhancement in strained Si p-MOSFETs was first demonstrated by Nayak et al. [11]. The initial devices were fabricated on a 1 μm thick uniform-composition partially relaxed SiGe buffer, which is known to have a very high defect density, which resulted in a limited performance (subthreshold slope 111 mV/decade). An improved device structure and process to fabricate high-performance strained Si p-MOSFETs with a step-graded completely relaxed thick (3 μm) SiGe buffer layer (defect density < 10^5 cm^{-2}), a low thermal budget (maximum temperature 700°C), and a high-quality (100 Å) gate oxide has been reported. As discussed earlier, strained Si is more difficult to grow than strained $Si_{1-x}Ge_x$, since bulk $Si_{1-x}Ge_x$ substrate is currently not available and, until recently, growth of relaxed $Si_{1-x}Ge_x$ without forming a large concentration of defects due to dislocations was difficult. Moreover, the thermal budget of a conventional CMOS process, which is largely dominated by the gate oxide growth and annealing process steps, needs to be minimised for the fabrication of strained Si/SiGe devices. Optimisation of the thermal budget is necessary because significant Ge outdiffusion and corresponding strain relaxation at process temperatures beyond 800°C are observed.

Problems arising from the dislocations include (1) reduction of carrier mobility due to scattering, which consequently lowers the operating speed of devices, (2) dopant diffusions along dislocation lines that cause current leakage, and (3) recombination of free carriers by trap states created from the dislocations, resulting in a lower current density of devices. However, it has been demonstrated experimentally that the hole mobility is improved in strained Si. The enhancement in the hole mobility was found to be 40% at room temperature and 200% at 77 K [12]. However, the ability to achieve both n- and p-MOSFET devices using strained Si provides a promising alternative for next generation high-performance SiGe CMOS technology.

The thermal conductivity of the underlying SiGe is at least 15 times lower than that of the bulk Si. Thus, in the case of a strained Si MOSFET, the SiGe layer basically confines the generated heat to the top layer of Si. This is analogous to the self-heating effect in silicon-on-insulator, and it will degrade the drive current. Due to self-heating, strain relaxation could also result, which may affect the drain current enhancement. From a reliability perspective, the degradation mechanism could be very sensitive to the transistor temperature. At high drain voltage, the heat is readily dissipated in the case of a bulk Si transistor, and hence there is a negligible difference between the drain current measured by the DC technique and that by the AC technique. In contrast, the drain current of the strained Si transistor measured by the DC technique is degraded under high drain voltage bias due to the low heat dissipation efficiency. As the AC measurement setup could relieve part of the heating process, some drain current enhancement would then be observed. The presence of self-heating effect is also confirmed by other measurement techniques. From the hot-carrier reliability perspective, the degradation mechanism could be very sensitive to the transistor temperature in the strained Si.

Strained Si quantum wells (QWs) on relaxed SiGe layers, generally known as relaxed buffer layers (also called virtual substrates), can be used for both n- and p-MOSFETs. Moreover, these virtual substrates are of great interest for integrating Si-based devices with III-V semiconductor devices to utilise their optical properties [1]. Very high electron mobilities demonstrated in strained Si layers suggest a great potential for this material in high-transconductance n-MOSFETs. To date, in-plane electron mobilities approaching 3,000 cm^2/Vs have been reported in long-channel MOSFETs with both surface and buried channels.

Due to its enhanced current drive and high-frequency performance, strained Si technology is undoubtedly one of the enabling technologies for RF circuit applications. Although the enhanced cutoff frequency of the strained Si MOSFET can facilitate the RF CMOS circuit design, the unintentionally induced threading dislocations in the strained Si channel can potentially degrade some RF circuits. Enhanced drive currents of 15 to 25% have been demonstrated on sub-100 nm bulk strained silicon MOSFETs. However, it has been difficult to implement because of misfit and threading dislocations, Ge outdiffusion, silicide formation difficulty, self-heating effect, higher arsenic diffusion in the S/D extension region, and cost.

2.10 Hybrid Orientation Technology

The electrical properties of Si are highly dependent on the crystal orientation and the direction of the carrier transport. Furthermore, hole and electron mobility are not maximised in the same direction and orientation. Besides substrate-induced and process-induced stress engineering, wafer substrate orientation and channel orientation can improve mobility. Different surface orientations and directions of applied field for different in-plane stresses provide different interactions with carrier transport. The strained Si technologies discussed so far assumed the standard (100) Si substrate for improving the carrier mobilities. Since the carrier mobilities are dependent on the crystal orientation of the substrate as well as the direction of the channel [13], the mobilities can be further enhanced by adopting different substrate orientations and channel directions. Maximum benefit in CMOS performance can be drawn when the n- and p-MOSFET transistors are grown on (100) and (110) substrate orientations, respectively, with [110] as the channel direction. Combining the benefits of this hybrid orientation technology (HOT) on SOI with the stress induced from processing steps, significant improvements in p-MOSFET mobility have been reported [14].

An alternative approach yielding mobility improvement in Si exploits the dependence of the carrier mobility in Si inversion layers on the crystal orientation and on the current flow direction. For example, for holes the mobility is 2.5 times higher for (110) surface orientation than for standard (001) orientation, depending on the applied effective vertical field. In HOT, which is based on wafer bonding techniques and selective epitaxy, the larger carrier mobility of holes for (110)-oriented substrate is exploited to enhance the performance of p-channel MOSFETs. HOT seems promising because processes are directly compatible with existing CMOS technology and strain engineering.

Yang et al. [14] have developed a novel planar silicon CMOS structure: HOT comprised of n-MOSFETs on silicon of (100) surface orientation and p-MOSFETs on (110) surface orientation (see Figure 2.19). A schematic cross section of CMOS on the hybrid orientation substrate is shown in Figure 2.20. HOT includes two types: type A with p-MOSFET on the (110) SOI and n-MOSFET on the (100) silicon epitaxial layer, and type B with n-MOSFET on the (100) SOI and p-MOSFET on the (110) silicon epitaxial layer.

The integrated process flow for the HOT CMOS fabrication is shown in Figure 2.21. It is found that electron mobility on the (100) epi-Si can be even slightly better than that on the (100) control substrate, and hole mobility on (110) epi-Si is 2.5 times that on the (100) control (Figure 2.22), similar to that observed on the (110) bulk substrate.

Figure 2.23 shows the drive current of large-width p-MOSFETs on (110) epitaxial silicon at –1.0 V supply voltage. p-MOSFET drive current is improved by 29% in the <110> direction from the compressive stress at I_{off} = 100 nA/μm, and little effect is observed when the current flow direction changes to <100>.

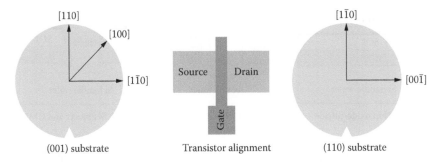

FIGURE 2.19
Crystallographic directions on the (001) and the (110) substrate. Conventionally, MOSFET channels are aligned along the <110> direction on the (001) substrate. Highest electron mobility is obtained on the (001) substrate, and highest hole mobility on the (110) substrate with channel direction <110>. (After Ungersboeck, S.-E., Advanced Modelling of Strained CMOS Technology, PhD thesis, Technical University, Wien, 2007.)

The combination of dual-stress liners (DSLs) with HOT provides excellent CMOS performance, with a record p-MOSFET drive current of 730 µA/µm at 90 nA/µm I_{off} and –1.0 V supply voltage. The subthreshold and output characteristics of p- and n-MOSFET are given in Figure 2.24. The subthreshold slope is 96 and 108 mV/dec, and drain-induced barrier lowering (DIBL) is 102 and 147 mV for p- and n-MOSFET, respectively.

2.10.1 Device Simulation

The hybrid orientation technology (HOT) combines different silicon substrate orientations and channel directions on the same wafer and can be used in conjunction with strain techniques. Since strain yields an anisotropic

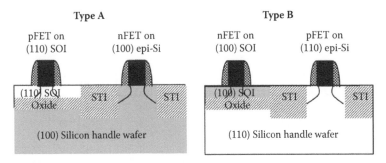

FIGURE 2.20
Schematic cross section of CMOS on hybrid orientation substrates, including two types: type A with p-MOSFET on (110) SOI and n-MOSFET on (100) silicon epitaxial layer, and type B with n-MOSFET on (100) SOI and p-MOSFET on (110) silicon epitaxial layer. (After Yang, M., V. W. C. Chan, K. K. Chan, L. Shi, D. M. Fried, J. H. Stahis, A. I. Chou, E. Gusev, J. A. Ott, L. E. Burns, and M. V. Fischetti, *IEEE Trans. Electron Dev.*, 53, 965–978, 2006. With permission.)

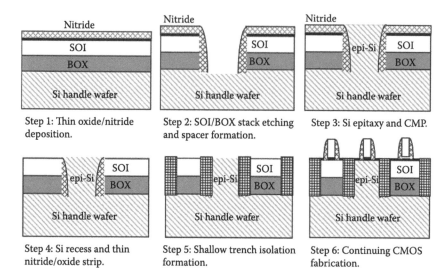

FIGURE 2.21
Integrated process flow for the HOT CMOS fabrication on a hybrid orientation substrate, where n-MOSFET is on the (100) surface and p-MOSFET is on the (110) surface. (After Yang, M., V. W. C. Chan, K. K. Chan, L. Shi, D. M. Fried, J. H. Stahis, A. I. Chou, E. Gusev, J. A. Ott, L. E. Burns, and M. V. Fischetti, *IEEE Trans. Electron Dev.*, 53, 965–978, 2006. With permission.)

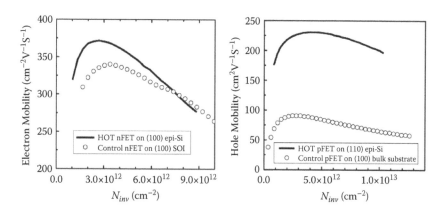

FIGURE 2.22
Electron (left) and hole (right) mobility measured on MOSFETs fabricated on (100) epi-Si on a HOT-A substrate and (110) epi-Si on a HOT-B substrate, respectively. (After Yang, M., V. W. C. Chan, K. K. Chan, L. Shi, D. M. Fried, J. H. Stahis, A. I. Chou, E. Gusev, J. A. Ott, L. E. Burns, and M. V. Fischetti, *IEEE Trans. Electron Dev.*, 53, 965–978, 2006. With permission.)

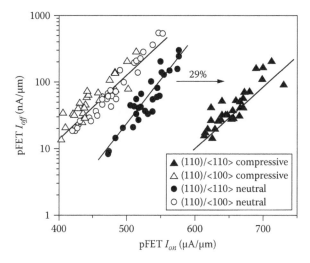

FIGURE 2.23
Characteristics of p-MOSFETs fabricated on (110) epi-Si on a HOT-B substrate with channel direction along <110> and <100> under uniaxial longitudinal compressive stress from a contact etch stop nitride liner (solid dots) compared with the stress-relaxed condition (open dots). (After Yang, M., V. W. C. Chan, K. K. Chan, L. Shi, D. M. Fried, J. H. Stahis, A. I. Chou, E. Gusev, J. A. Ott, L. E. Burns, and M. V. Fischetti, *IEEE Trans. Electron Dev.*, 53, 965–978, 2006. With permission.)

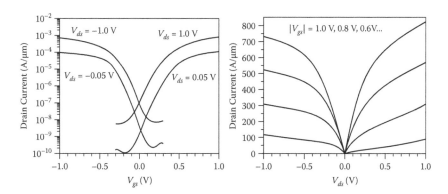

FIGURE 2.24
Subthreshold and output characteristics of HOT CMOS fabricated on a HOT-B substrate. DSLs were implemented on n-MOSFETs and p-MOSFETs to improve performance. (After Yang, M., V. W. C. Chan, K. K. Chan, L. Shi, D. M. Fried, J. H. Stahis, A. I. Chou, E. Gusev, J. A. Ott, L. E. Burns, and M. V. Fischetti, *IEEE Trans. Electron Dev.*, 53, 965–978, 2006. With permission.)

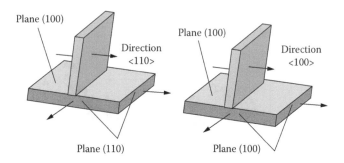

FIGURE 2.25
Representation of transistors in three dimensions with the sides of active having a surface of substrate (100). In this case: left-hand side with an orientation <110> and the sides (110) and right-hand side with an orientation <100> and the sides (100). (After Maiti, T. K., Process-Induced Stress Engineering in Silicon CMOS Technology, PhD thesis, Jadavpur University, 2009.)

mobility, the proper channel direction and substrate orientation have to be chosen to obtain the maximum mobility enhancement. In the hybrid orientation technology, which is based on wafer bonding techniques and selective epitaxy, the larger carrier mobility of holes for (110)-oriented substrate is exploited to enhance the performance of p-MOSFETs. HOT seems promising because processes are directly compatible with existing CMOS technology and strain engineering. Let us consider the architecture of the transistor in three dimensions: devices being insulated by trenches; the sides of the active zones are not directed in the same way according to the substrate. In our cases when we have a substrate directed <110>, with a flow of current in the transistors in the direction <110>, the sides of active for the insulation are then plans (110). In the case <100>, the sides are thus plans (100) (Figure 2.25). Simulated strain-engineered MOS devices with hybrid orientation technology have been presented in Figure 2.26.

Therefore, a large amount of research today is focused on achieving substrates with the so-called hybrid orientation technique (HOT). In this approach, the n- and p-MOSFETs are processed for (100) and (110) Si crystal orientations, respectively. The main advantage with this technology is that no novel material is introduced. Therefore, normally only minor changes to the processing sequence are anticipated (e.g., channeling differences during implantation and more complex substrate manufacturing). However, recent results point out that difficulties arise when SiGe is introduced on a surface other than (100) due to relaxation and faceting. The advantages of SOI and biaxial strained Si layers can be combined in a single substrate of strained silicon-on-insulator (SSOI).

RF performance of process-induced strained Si p-MOSFETs in hybrid orientation technology has also been studied using technology CAD tools [15] that properly account for the physical mechanisms, such as orientation-dependent and process-induced strain-dependent mobility models. The

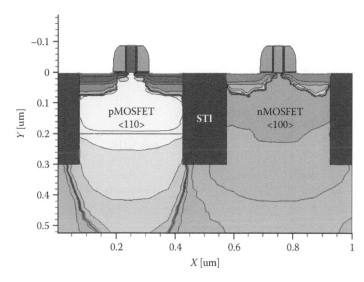

FIGURE 2.26
Simulated strain-engineered MOS devices with hybrid orientation technology (HOT). (After Maiti, T. K., Process-Induced Stress Engineering in Silicon CMOS Technology, PhD thesis, Jadavpur University, 2009.)

$Si_{0.83}Ge_{0.17}$ pockets in the source and drain region induce compressive stresses in different areas of the structure, including the channel. The effects of mobility enhancement, induced by surface orientation change and also process-induced strain, simultaneously, on the RF performance of p-MOS-FETs are taken into account. The frequency response of the e-SiGe MOSFETs was simulated in the common source configuration with f_T as the unity gain cutoff frequency. A cutoff frequency, f_T, of about 240 GHz is predicted for p-MOSFETs, in hybrid orientation technology involving process-induced strain. It is found that the f_T is higher in the <110> direction than in the <100> direction. This result indicates the advantage of strain-dependent mobility enhancement along with hybrid orientation toward high-speed device design. HOT seems promising because processes are directly compatible with existing CMOS technology and strain engineering. The above discussion illustrates the benefits of using hybrid crystal orientation substrates for significantly improving p-MOSFET performance. An obvious extension of this approach is to implement alternative crystalline orientations of novel channel materials, such as Ge. Epitaxial growth of Ge has been attempted on bulk (110) Si wafers.

The advantages of both the (110) orientation and Ge for higher mobility can be leveraged to attain a three times enhancement in the hole mobility compared to universal Si/SiO_2 hole mobility. Comparison of relative enhancement in low-field-hole mobility achieved by a combination of mobility enhancement techniques, i.e., the use of new materials and alternative crystal

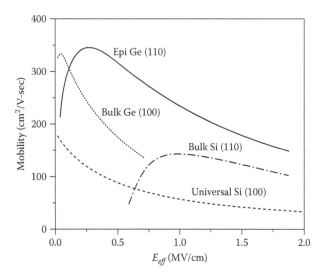

FIGURE 2.27
Comparison orientation dependence of low-field-hole mobility enhancement in Si and Ge. (After Joshi, S. V., Novel Channel Materials for Si-Based MOS Devices: Ge, Strained Si, and Hybrid Crystal Orientations, PhD dissertation, University of Texas at Austin, May 2007.)

orientations, shows significant promise [16]. Figure 2.27 shows enhancements in hole mobility that could be achieved by concomitant implementation of two mobility enhancement techniques—the use of Ge for its intrinsically higher mobility, and the (110) surface for additional hole mobility improvements. For comparison purposes, mobility data from the bulk Ge (100) and bulk Si (110) devices with a poly-Si/SiON gate stack are also plotted.

2.11 Summary

The focus of this chapter has been the application of SiGe, SiGe:C, strained Si, Ge, and strained Ge to MOSFET devices, with the goal of determining what enhancements in device performance can be achieved and understanding the fabrication complexity. Enhancing the carrier mobility in the channel of a Si MOSFET has the potential to extend the performance limits of existing CMOS technology. Theoretical calculations indicate that inducing strain in Si will split the degeneracy at the conduction and valence band minima, since relaxed silicon-germanium ($Si_{1-x}Ge_x$) has a larger lattice constant than bulk Si. Thin Si layers grown pseudomorphically on this material will be strained in biaxial tension. This produces enhanced in-plane carrier mobility in the strained Si as well as band offsets between the relaxed $Si_{1-x}Ge_x$ and the strained Si, which can be used for carrier confinement in advanced device structures. Surface channel strained Si p- and n-MOSFETs with varying strain in the Si layer were discussed.

A variety of strained Si on relaxed $Si_{1-x}Ge_x$ heterostructure architectures for p- and n-MOSFETs have been considered in order to study the carrier transport. The nature of the carrier mobility enhancement in the strained Si n- and p-MOSFETs has been discussed. A broad range of experimental techniques used for the growth of the films, the effect of various fabrication processes, and the electronic properties of the resulting structures have been covered. In addition to clearly demonstrating enhanced device performance in long-channel MOSFETs, several important physical characteristics of the strained Si material itself were also discussed.

Review Questions

1. What is substrate-induced strain?
2. What are the effects of strain on mobility?
3. Electron mobility is increased by the degeneracy splitting of the conduction band minimum. (True/False)
4. MOSFET performance is degraded at high vertical electric fields. (True/False)
5. What is biaxial strain?
6. What is uniaxial strain?
7. What are the differences between biaxial and uniaxial strain?
8. How can strain be introduced in a semiconductor?
9. What are the two main strain technologies being used in CMOS fabrication?
10. What is hybrid orientation technology?
11. What is process-induced strain?
12. What is critical layer thickness?
13. What is band gap engineering?
14. What is Vegard's rule?
15. What happens when one deposits film beyond a critical layer thickness?
16. What is SSOI?
17. What is the use of the relaxed SiGe buffer layer?

References

1. C. K. Maiti and G. A. Armstrong, *Applications of Silicon-Germanium Heterostructure Devices*, Institute of Physics Publishing (IOP), London, UK, 2001.
2. C. K. Maiti, N. B. Chakrabarti, and S. K. Ray, *Silicon Heterostructures: Materials and Devices*, Institute of Electrical Engineers (IEE), London, UK, 2001.
3. B. S. Meyerson, UHV/CVD Growth of Si and Si-Ge Alloys: Chemistry, Physics, and Device Applications, *Proc. IEEE*, 80, 1592–1608, 1992.
4. C. K. Maiti, Editorial in Special Issue on Strained-Si Heterostructures and Devices, *Solid-State Electronics*, 48, 1255, 2004 (and references there in).
5. C. K. Maiti, S. Chattopadhyay, and L. K. Bera, *Strained-Si Heterostructure Field Effect Devices*, CRC Press, Boca Raton, FL, 2007.
6. C. O. Chui, H. Kim, D. Chi, B. B. Triplett, P. C. McIntyre, and K. C. Saraswat, A Sub-400°C Germanium MOSFET Technology with High-k Dielectric and Metal Gate, *IEEE IEDM Tech. Dig.*, 437–440, 2002.
7. C. Claeys and E. Simoen, *Germanium-Based Technologies: From Materials to Devices*, Elsevier, Amsterdam, 2007.
8. H. Zang, W. Y. Loh, J. D. Ye, G. Q. Lo, and B. J. Cho, Tensile-Strained Germanium CMOS Integration on Silicon, *IEEE Electron Dev. Lett.*, 28, 1117–1119, 2007.
9. M. Y. A. Yousif, Silicon-Germanium for High-Performance CMOS Technology, PhD thesis, Chalmers University of Technology and Goteborg University, 2001.
10. G. S. Kar, S. Maikap, S. K. Banerjee, and S. K. Ray, Series Resistance and Mobility Degradation Factor in C-Incorporated SiGe Heterostructure p-Type Metal–Oxide Semiconductor Field-Effect Transistors, *Semicond. Sci. Technol.*, 17, 938–941, 2002.
11. D. K. Nayak, K. Goto, A. Yutani, J. Murota, and Y. Shiraki, High-Mobility Strained-Si PMOSFETs, *IEEE Trans. Electron Dev.*, 43, 1709–1715, 1996.
12. C. K. Maiti, L. K. Bera, S. S. Dey, D. K. Nayak, and N. B. Chakrabarti, Hole Mobility Enhancement in Strained-Si p-MOSFETs under High Vertical Fields, *Solid-State Electron.*, 41, 1863–1869, 1997.
13. H. S. Momose, T. O. K. Kojima, S. Nakamura, and Y. Toyoshima, 110 GHz Cutoff Frequency of Ultra-Thin Gate Oxide P-MOSFETs on [110] Surface-Oriented Si Substrate, *VLSI Symp. Tech. Dig.*, 156–157, 2002.
14. M. Yang, V. W. C. Chan, K. K. Chan, L. Shi, D. M. Fried, J. H. Stahis, A. I. Chou, E. Gusev, J. A. Ott, L. E. Burns, and M. V. Fischetti, Hybrid-Orientation Technology (HOT): Opportunities and Challenges, *IEEE Trans. Electron Dev.*, 53, 965–978, 2006.
15. T. K. Maiti, S. S. Mahato, S. K. Sarkar, and C. K. Maiti, Performance Enhancement of p-MOSFETs with Embedded SiGe Source/Drain on Hybrid Orientation Substrates, *Proc. ULIS 2007*, 2–6, 2007.
16. S. V. Joshi, Novel Channel Materials for Si Based MOS Devices: Ge, Strained Si and Hybrid Crystal Orientations, PhD dissertation, University of Texas at Austin, May 2007.
17. P. Zimmerman, High Performance Ge pMOS Devices Using a Si-Compatible Process Flow, *IEEE IEDM Tech. Dig.*, 261–264, 2006.

3

Process-Induced Stress Engineering in CMOS Technology

For more than four decades the rapid progress in complementary metal-oxide-semiconductor (CMOS) technology has taken place through the tremendous pace of scaling, leading to an enormous increase in speed and functionality of electronic devices. However, it is getting extremely difficult to meet metal-oxide-semiconductor field-effect transistor (MOSFET) performance gains with acceptable device leakage. Now the gate leakage current constitutes a major part of the power budget of microprocessors. Another critical scaling issue involved is the increase of the source/drain series resistance resulting from the ultra-shallow p-n junctions in the source/drain regions. To keep the source/drain series resistance at a reasonable fraction of the total channel resistance (~10%), several alternative MOSFET structures have been proposed, such as nonoverlapped gate structures, which do not require ultra-shallow source/drain junctions or structures with metallic source and drain electrodes to minimise the series resistance. Advanced multigate structures, such as FinFETs and ultra-thin-body (UTB) MOSFETs, may provide a path toward scaling CMOS to the end of the ITRS road map. Stress and strain engineering are the key elements in current CMOS technologies and can accommodate nonclassical CMOS structures.

Starting at the 65 nm node, stress engineering to improve performance of transistors has been a major industry focus. In order to induce appropriate strain in the channel region of MOSFETs, various techniques have been introduced, such as substrate-induced strain, process-induced strain, and bending-induced strain. The epitaxially grown Si on a relaxed $Si_{1-x}Ge_x$ layer is a typical example of substrate-induced strain. The lattice of the Si layer is stretched (biaxial tensile strain) in the plane of the interface due to the lattice mismatch between Si and $Si_{1-x}Ge_x$. By increasing Ge mole fraction (x), more biaxial tensile strain in the Si layer is induced as long as its thickness is under critical thickness. However, strain relaxation during high-temperature processes and high defect density (e.g., misfit and threading dislocations) remain issues for production. In addition, the hole mobility enhancement is reduced at high electric field for biaxial tensile strain. Recent attention has shifted to process-induced uniaxial strain as uniaxial compressive strain along a <110> channel enhances hole mobility even at high vertical fields. Strain is one key feature to enhance the performance of Si MOSFETs. Biaxial tensile strain has been investigated both experimentally and theoretically in CMOS technology [1]. It improves the electron mobility, but degrades the hole mobility at low-stress range (<500 MPa). Since 2003, uniaxial stress has been applied to Intel's 90, 65, 45, and 22 nm technologies to improve the drive current without significantly increased manufacturing complexity [2, 3]. Advanced CMOS technologies feature multiple process-induced stressors such as compressive and tensile overlayers, embedded SiGe (e-SiGe), and multiple stress memorisation techniques. Large magnitudes of uniaxial channel stress (~1 GPa) are being incorporated in p-channel devices of the 65 nm technology node, and an even higher stress level is necessary in the 32 nm technology node. Local strain approaches are based on dedicated processing steps or process modules such as shallow trench isolation, silicidation or metal gate electrodes, the use of liners and capping layers, dry etch processes, contact etch stop layers, and source/drain (S/D) engineering.

Extension of CMOS beyond 22 nm technology nodes will require new nonclassical MOSFET structures coupled with advanced materials and processes. Classes of new materials include high-k gate dielectrics, metal and mid-gap gate metal electrodes, strained Si, and silicon-germanium alloys. These new materials will lower the gate leakage current and gate resistance, reduce the poly-gate electrode depletion capacitance, and increase the device speed. Nonclassical CMOS structures offer better control of short-channel effects, improved ON current via higher channel mobility, lower load capacitance, and lower propagation delay time. Besides scaling, several innovative mobility enhancement techniques are being attempted to maintain the CMOS performance improvement. Mobility enhancement is attractive because it improves device performance without device scaling. However, continued miniaturisation increases device complexity and internal mechanical stress.

Changes in electron and hole mobility due to stress from local oxidation of silicon (LOCOS) and shallow trench isolation have been known for a long time. But because the strain from a localised source decays rapidly away from the stressor, it could not be used as a strain technique until deep submicrometer technologies were developed. Stress or strain changes the band structure of a semiconductor, which in turn changes other material properties, such as band gap, effective mass, carrier mobility, diffusivity of dopants, and oxidation rates. When applied in the direction of the channel (for standard wafer orientation), tensile strain is used to improve the electron mobility in n-MOSFETs, while compressive strain is beneficial for hole mobility improvement in p-MOSFETs.

One example of the local strain approach is to integrate epitaxially grown SiGe into the source and drain regions. A compressive stress in the direction of the device channel can be generated, if SiGe with its larger lattice constant is grown epitaxially on silicon. However, this approach can only be used for enhancement of p-MOSFET devices. For n-MOSFET devices, a similar effect can be achieved by epitaxially growing a material with a smaller crystal lattice, such as carbon-doped silicon (SiC), into the source and drain regions.

The technique for local strain introduction for both n- and p-MOSFET devices is the contact etch stop layer (CESL). It consists of a nitride liner, deposited over the devices. By tuning the process parameters, the type and level of the intrinsic stress of the liner can be determined. The so-called dual-CESL approach consists of depositing a CESL liner with tensile stress over the n-MOSFET devices and a liner with compressive stress over the p-MOSFET devices, thus improving both electron and hole mobility at the same time. This makes the dual-CESL approach one of the leading candidates for the CMOS industry.

Major techniques to introduce uniaxial stress include embedded SiGe (e-SiGe) technology, dual-stress liner (DSL), stress memorisation technique (SMT), and the parasitic stress from shallow trench isolation (STI). Embedded SiGe in the source and drain area is used to introduce compressive stress for p-MOSFET. DSL introduces the stress by depositing a highly stressed silicon nitride layer, tensile stress for the n-MOSFET region, and compressive stress for the p-MOSFET region, over the entire wafer to elevate carrier mobility. In SMT, the stress in the channel is transferred from the stressed deposited dielectric and is memorised during the recrystallisation of the active area and poly-gate when thermal annealing is activated. STI stress results from the difference in thermal expansion coefficients between SiO_2 and Si. It is an intrinsic stress source and not intentionally built up for enhancing device performance enhancement. The purpose of this chapter is to briefly review the currently used promising strain techniques to fabricate strained silicon transistors, and to assess their opportunities, as well as their technological limitations. Strain induced by epitaxial $Si_{1-x}Ge_x$ in the source/drain regions and strained contact etch stop layers (CESLs) are covered. Layout dependence of the $Si_{1-x}Ge_x$ S/D and strained CESL technologies are discussed.

3.1 Stress Engineering

An engineered substrate is a semiconductor material that can be fabricated and introduced in the conventional silicon manufacturing, resulting in products that are unique and could not have been fabricated using only silicon substrate. The introduction of strain changes the mechanical, electrical (band structure and mobility), and chemical (diffusion and activation) properties of a semiconductor. The various effects of stress and strain on silicon and also silicon technology have been studied since 1950s [4, 5]. Most significant to silicon technology are the changes in band gap, effective mass, mobility, diffusivity of dopants, and oxidation rates. The effects of strain on mobility were found to be anisotropic, and carrier effects were found to be different for bulk silicon and inversion layers [6, 7].

Classification of strain techniques currently in use can be made in two main categories. Strain is introduced across the entire substrate in global

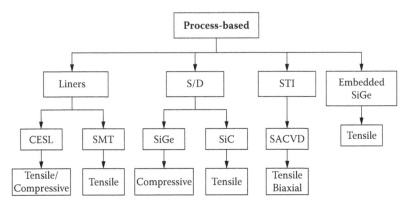

FIGURE 3.1
Different process-based mobility enhancement technologies. (After Hallstedt, J., Epitaxy and Characterization of SiGe:C Layers Grown by Reduced Pressure Chemical Vapor Deposition, PhD thesis, Royal Institute of Technology (KTH), 2004.)

strain techniques, whereas local techniques induce strain in selected regions of the wafer. Some of the most prominent strain technologies that are currently in use in industry are shown in Figure 3.1. These mobility boosters are usually effective only at gate lengths below roughly 100 nm. A key challenge of all technologies is their ability to be integrated into the CMOS manufacturing process and to avoid an increase in processing costs. Process-induced strain techniques are generally not universal in their implementation and need to be tailored to a particular transistor integration scheme. The local strain approach has currently turned out to be more promising in CMOS technology and is the first strain technology used in high-volume production. Table 3.1 shows major sources of process-induced stress to enhance MOSFET performance. Tensile stress is used for n-MOSFETs to induce tensile localised strain to improve electron mobility. The stress memorisation technique (SMT) is also used to improve n-MOSFET performance.

Process-induced strain can be applied during the fabrication process by adding new process steps or using existing process steps with relatively

TABLE 3.1

Main Techniques Used for Process-Induced Stress Generation

Process-Induced Stress Using	Improves
Single stress liner	n-MOSFET or p-MOSFET
Embedded SiGe in S/D (e-SiGe)	p-MOSFET
Stress memorisation technique (SMT)	n-MOSFET
Dual-stress liner (DSL)	n-MOSFET and p-MOSFET
DSL + e-SiGe	n-MOSFET and p-MOSFET
Stress proximity technique (SPT) for DSL	n-MOSFET and p-MOSFET

low cost. In addition, typical process-induced strain technologies, such as strained capping layer and embedded $Si_{1-x}Ge_x$ source/drain, are known to generate uniaxial strain along the channel, which offers similar electron mobility enhancement compared with biaxial strain, while the hole mobility enhancement is retained at high E_{eff}. The integration of SMT/DSL/e-SiGe has been demonstrated in the literature for high-performance CMOS. Recently, the stress proximity technique (SPT) for DSL has been successfully demonstrated by removing the spacer between the stressed liner and poly-gate to maximise stress proximity. The integration of cost-effective techniques of SMT/DSL/SPT has also been demonstrated. The various sources of process-induced stress are briefly described below.

3.2 $Si_{1-x}Ge_x$ in Source/Drain

Local strain in the device channel can be induced by substituting the Si in the source/drain regions by a material with a different lattice constant. This technique was first proposed by Intel as a strain technique for performance enhancement. SiGe has been used in the past in the source and drain regions for higher boron activation and reduced external resistance. Interestingly, embedded SiGe at the source and drain has been recognised as one of the options offering the best potential to enhance performance in sub-100 nm technologies. This is based on a two-step process to form recessed junctions: a dry or vapour etch of Si in S/D regions and a selective epitaxy growth of B-doped SiGe layers (inducing compressive strain). The improvement of the transistor performance is not only a mobility enhancement but also from reduced S/D access resistance. Use of a lattice mismatched S/D stressor, such as $Si_{1-x}Ge_x$ in the S/D region, is a very promising technique to introduce local compressive strain in the channel region for p-MOSFET performance boost. More than 50% enhancement in hole mobility over universal mobility can be achieved with only a few key process steps added to the standard CMOS fabrication. In this case, strain is introduced from the stressor side using $Si_{1-x}Ge_x$ in the source and drain region of the p-MOSFETs and $Si_{1-x}C_x$ for n-MOSFETs. The $Si_{1-x}Ge_x$ in the source and drain process flow is simple, low in cost, and solves the major issues associated with the biaxial strained silicon on the relaxed $Si_{1-x}Ge_x$ buffer layer approach. Figure 3.2 shows the beneficial strain components desired for improving hole mobility in a p-channel silicon transistor having a (001) channel surface and a source-to-drain direction oriented along a [110] crystal direction.

Since in p-MOSFET devices, compressive parallel stress is beneficial for hole mobility improvement, the Si in the S/D regions is substituted by $Si_{1-x}Ge_x$ (see Figure 3.3). The larger lattice constant of $Si_{1-x}Ge_x$ creates a compressive parallel stress inside the channel. The lattice constant of the

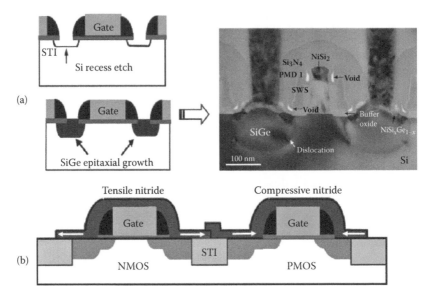

FIGURE 3.2
(a) Strained Si p-channel MOSFET process flow for the representative stacked gate transistor and transmission electron microscopy cross-sectional view. (b) Dual-stress liner process architecture with tensile and compressive silicon nitride capping layers. (After Thompson, S. E., G. Sun, Y. Choi, and T. Nishida, *IEEE Trans. Electron Dev.*, 53, 1010–1020, 2006. With permission.)

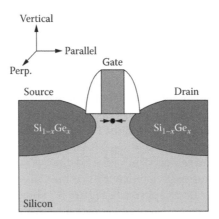

FIGURE 3.3
Cross section of a PMOS device with embedded $Si_{1-x}Ge_x$ source/drain regions. The bigger lattice of the $Si_{1-x}Ge_x$ alloy results in compressive parallel stress in the channel of the device. (After Eneman, G., Design, Fabrication, and Characterization of Advanced Field Effect Transistors with Strained Silicon Channels, PhD thesis, Katholieke Universiteit Leuven, 2006.)

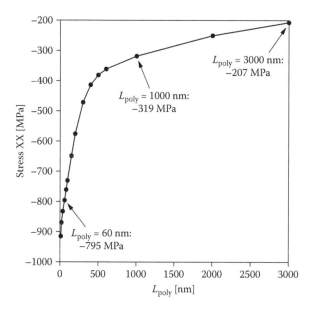

FIGURE 3.4

Parallel stress dependence on channel length. This dependence makes the $Si_{1-x}Ge_x$ S/D technique very promising in terms of scalability. (After Shickova, A., Bias Temperature Instability Effects in Devices with Fully-Silicided Gate Stacks, Strained-Si, and Multiple-Gate Architectures, PhD thesis, Katholieke Universiteit-Leuven, 2008.)

$Si_{1-x}Ge_x$ alloy depends on the Ge concentration. Consequently, the stress in the channel also depends on the Ge concentration. The strain in the channel also depends on the source/drain regions etch depth and on the $Si_{1-x}Ge_x$ overgrowth. Typically, germanium concentrations around 20% are used for the S/D regions. Ge concentrations above 30% are not beneficial due to increased defect formation and subsequent strain relaxation. The stress in the channel increases with decreasing channel length, and a compressive stress in the order of 1 GPa can be achieved for gate lengths of ~50 nm (see Figure 3.4). The channel length dependence makes the $Si_{1-x}Ge_x$ S/D technique very promising in terms of scalability.

By confining the $Si_{1-x}Ge_x$ to the source and drain and introducing it late in the process flow, yield issues with threading dislocations are eliminated, midsection thermal cycles are unaltered, significantly thinner $Si_{1-x}Ge_x$ is needed, source and drain extensions are still formed in silicon as opposed to $Si_{1-x}Ge_x$, and self-heating caused by low thermal conductivity of the $Si_{1-x}Ge_x$ in the substrate is unchanged. The origin of the strain in the channel region is from the interaction between the pair of lattice-mismatched materials at the semiconductor heterojunction, which induces lateral compressive strain along the Si channel direction and enhances the hole mobility. Verheyen et al. [8] have reported a current enhancement of 25% over control

FinFETs. Unlike the planar device, the major conducting surfaces of the FinFET devices came from the sidewalls of the fin, having a surface orientation of (110). To exert a strain in the channel similar to that of the planar 2D devices, an embedded SiGe around the fin at the S/D regions is needed. Local Ge condensation has been reported as an alternative technique for the fabrication of embedded SiGe S/D stressors. It is reported that during the oxidation of the SiGe film, Ge is rejected from the oxide, and this caused a pile-up of Ge at the interface between the top SiO_2 layer and SiGe. Compressive stress is exerted on the channel from the SiGe S/D regions on both the top and sidewalls. FinFETs with condensed SiGe S/D show a higher drive current.

Two- and three-dimensional finite element (FE) methods are used to study the stress and strain in the transistor structure. Lattice spacing of this $Si_{1-x}Ge_x$ material is larger than silicon and results in uniaxial compressive strain in the channel region, as shown in Figure 3.3. For p-MOSFET with $Si_{1-x}Ge_x$ S/D, numerical simulation studies indicate that the magnitude of the lateral compressive strain and the vertical tensile strain induced in the Si channel can be increased by increasing the Ge mole fraction x in the $Si_{1-x}Ge_x$ S/D region, by increasing the recess depth of the $Si_{1-x}Ge_x$ S/D, or by reducing the separation between the $Si_{1-x}Ge_x$ S/D regions. The compressive stress is mainly dependent on the e-SiGe thickness in the S/D, both below and above (raised S/D) the Si surface, and Ge content in the SiGe. Too high a Ge content may cause defects and yield becomes an issue. Figure 3.5 shows the simulated XX component of stress tensor (σ_{xx}) of p-MOSFETs in channel with $Si_{0.83}Ge_{0.17}$ pockets. For p-channel transistors, the SiGe source and drain (S/D) stressor is commonly used to induce strain in the device channel. By making use of the lattice mismatch between Si and SiGe, compressive strain can be induced in the channel to enhance hole mobility. For enhanced strain effect, a recess etch can be performed on the S/D region prior to the SiGe epitaxial growth to realise embedded SiGe S/D stressors. Selective growth of SiC in the source and drain has been used for n-MOSFET devices, and it is similar to p-MOSFET with SiGe, but the recessed S/D is filled with SiC (inducing tensile strain). High C content and finding an optimised junction depth are two important issues to obtain a maximum electron mobility enhancement. The growth is somewhat more complicated due to low growth rates as a result of chlorine-based chemistry to preserve the selectivity mode. The process-induced uniaxial compressive stress leads to drive current improvements of up to 35% for PMOS transistors, which offers greater device performance. Figure 3.6 shows a cross section of a transmission electron microscopy (TEM) image of a PMOS test structure featuring an embedded SiGe source and drain. For 20% Ge, compressive channel stresses on the order of ~1 GPa are induced, depending on the proximity of the SiGe to the channel.

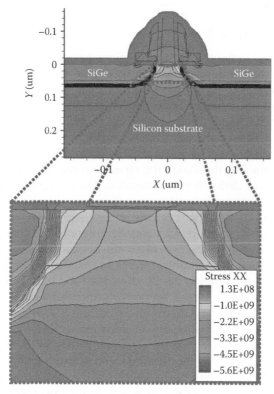

FIGURE 3.5
Simulated XX component of stress tensor (σ_{xx}) of p-MOSFETs in channel with $Si_{0.83}Ge_{0.17}$ pockets. (After Maiti, T. K., Process-Induced Stress Engineering in Silicon CMOS Technology, PhD thesis, Jadavpur University, 2009.)

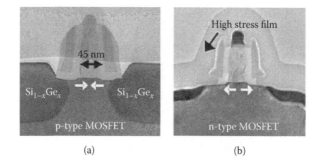

FIGURE 3.6
Uniaxillay and biaxially strained Si MOSFETs. (a) p-MOSFET with $Si_{1-x}Ge_x$ source and drain. (b) Nitride capping-layered n-MOSFET. (After Sun, G., Strain Effects on Hole Mobility of Silicon and Germanium p-Type Metal-Oxide-Semiconductor Field-Effect-Transistors, PhD thesis, University of Florida, 2007.)

3.3 Si$_{1-y}$C$_y$ in Source/Drain

SiC has been used in the source/drain regions to introduce stress locally for device drive current enhancement. Figure 3.7 shows the transistor structure with Si$_{1-y}$C$_y$ stressors in the source and drain regions. The SiC regions act as stressors, giving rise to lateral tensile strain and vertical compressive strain in the channel to enhance electron mobility. In addition, the SiC/strained Si heterojunction at the source end of the transistor provides for enhanced electron injection velocity from the source. The theoretical limit of the channel stress is determined by the maximum stress that can be generated at the Si/SiC interface before dislocation is generated. The maximum stress depends directly on the carbon mole fraction used. The partially relaxed SiC stressors in the source and drain regions tensile strain the Si channel laterally, leading to a large tensile stress that extends throughout the channel region. It has been observed that in the case of an anisotropic recess etch, for a given carbon mole fraction the amount of stress in the channel is determined by etch depth, which correlates to the SiC thickness, and the etch shape. The SiC stressors affect two major strain components, the lateral stress and the vertical stress. The magnitude and distribution of stress components, the origin of the stress field, and their relationship to electron mobility enhancement have been discussed. It is shown that the strain effect due to the SiC S/D stressors as well as the increased electron injection velocity may play an important role at sub-100 nm gate lengths. Reducing the interstressor spacing and increasing the C content and the recessed depth/raised height of the SiC stressors are three ways to achieve high strain levels in the Si channel region

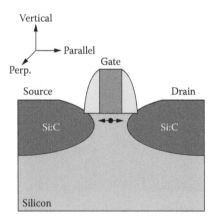

FIGURE 3.7
Cross section of an n-MOSFET device with embedded SiC source/drain regions. The smaller lattice of the SiC alloy results in tensile parallel stress in the channel of the device (After Eneman, G., Design, Fabrication, and Characterization of Advanced Field Effect Transistors with Strained Silicon Channels, PhD thesis, Katholieke Universiteit Leuven, 2006.)

for drive current and enhanced electron mobility in n-MOSFETs. To give better device performance, SiGe or SiC, which is lattice mismatched to Si, can be grown instead to induce strain in the channel for mobility improvement. In addition, SiGe can also be exploited to lower the contact resistance due to the smaller band gap of SiGe as Ge concentration increases.

3.4 Shallow Trench Isolation (STI)

Another technique, shallow trench isolation (STI), is normally used for lateral isolation between devices on the Si substrate. Shallow trench isolation is an important and well-studied stress source that has not been fully exploited until now for design quality improvement. STI usually exerts a compressive stress along the channel (i.e., the current flow direction), which improves p-MOSFET device mobility. The opposite type of stress, tensile stress, degrades the p-MOSFET performance in this direction. The STI etch process is used to create shallow trenches in the Si substrate, which are subsequently filled with dielectric material to form isolation barriers between the STI edge and the transistor region, creating a comprehensive channel between two trench structures. Routine CMOS processing operations, such as isolation formation, induce some strain in the silicon lattice. At STI, a large volume of oxide is deposited in a trench to create isolation structures. The volume contraction between the silicon and oxide results in stress from different coefficients of thermal expansion. Residual stress can also arise if the wafer is quenched (rapidly cooled), effectively locking a higher stress state oxide. Further, densification of the oxide at high temperatures due to changes in bonding can build up intrinsic residual stress in the STI oxides. This residual stress is compressive for most commonly used oxide filling materials [9].

Since n-MOSFET performance improves in the presence of tensile strain, recent efforts have focused on developing an STI process that results in tensile strain in the channel. The strains created at the isolation edge decay monotonically toward the middle of the channel, and the distance between the gate edge and isolation edge determines the actual strain in the channel. The larger the active area, the higher will be the impact of the silicide stress, while the narrow-width devices show higher influence of the STI stress. The magnitude of strain induced from the STI is typically lower than other forms of deliberate strain introduction. The popularly used BSIM SPICE model (revision 4.3 and higher) contains an explicit STI model. However, only the impact of the distance from the transistor channel to the STI boundary is modeled. Hence, the dependency on the STI width is not present in the BSIM4 model. Our simulations, as well as simulations and data in the literature [10], show that STIW impact cannot be neglected. Thus, as noted above, our present work not only models

STIW impact, but also builds upon this modelling to improve circuit performance at no area cost.

3.5 Contact Etch Stop Layer (CESL)

The contact etch stop layer technology is a local strain introduction technique alternative to the $Si_{1-x}Ge_x$ S/D and SiC S/D epitaxial growth techniques. The CESL technology exploits the intrinsic strain of the nitride contact etch liners (Figure 3.8). A tensile intrinsic stress in the CESL results in a tensile parallel stress in the channel and a compressive vertical stress, while the perpendicular stress is negligible. Since a tensile parallel stress and a compressive vertical stress are favourable for electron mobility, tensile CESL is ideal for n-MOSFET performance improvement. Similarly, a compressive intrinsic stress in the CESL leads to a compressive parallel channel stress and a tensile vertical stress. Therefore, it is beneficial for p-MOSFET devices. For hole mobility improvement only the parallel stress induced by the CESL is important, because hole conduction is insensitive to vertical stress. Thicker CESL leads to higher channel stress, but the stress starts to saturate for CESL thicknesses above 40–50 nm.

The above method cannot provide performance improvement for both n- and p-MOSFET devices. For example, if only a tensile liner is deposited, it will be beneficial for the n-MOSFET devices but detrimental for the PMOS devices. In order to achieve ultimate CMOS performance the dual-CESL approach has

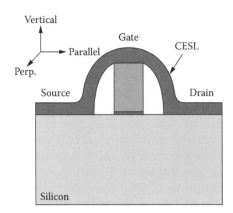

FIGURE 3.8
Cross section of a device with a deposited contact etch stop layer. The CESL can have either tensile or compressive intrinsic stress. (After Eneman, G., Design, Fabrication, and Characterization of Advanced Field Effect Transistors with Strained Silicon Channels, PhD thesis, Katholieke Universiteit Leuven, 2006.)

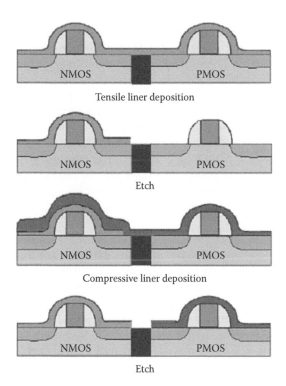

FIGURE 3.9
Simplified schematic diagrams of the main process flow steps for dual-CESL technology. A tensile liner is deposited on top of the n-MOSFET and a compressive liner on top of the p-MOSFET devices. (After Shickova, A., Bias Temperature Instability Effects in Devices with Fully-Silicided Gate Stacks, Strained-Si, and Multiple-Gate Architectures, PhD thesis, Katholieke Universiteit Leuven, 2008.)

been developed. A simplified representation of the main process flow steps for dual-CESL deposition is shown in Figure 3.9. The main steps are

1. Depositing a tensile liner
2. Performing a lithography step to define the liner on the NMOS
3. Removing the etch liner from the PMOS
4. Post-etch strip
5. Depositing a compressive CESL liner
6. Performing a lithography step to define the compressive liner on the PMOS
7. Removing the compressive liner from the NMOS
8. Post-etch strip

Stressed liner technologies improve transistor mobility by depositing a stressed nitride liner instead of a neutral liner on top of the gate and spacers of a transistor. Depending on the technology, either a compressive liner on top of p-MOSFET devices or a tensile liner on top of n-MOSFET devices can be used, with the remaining device type having a neutral liner on the top. The liner can be neutralised by doping. Dual-stress liner technology, on the other hand, uses both types of stressed liners and targets to improve both n- and p-MOSFET mobilities at the same time. Looking at the layout view, a p-MOSFET transistor is present on the left and an n-MOSFET transistor is present on the right. In the side view, we can see that p-MOSFET is covered with a compressive liner and n-MOSFET with a tensile liner. The boundaries of these liners are defined by the dashed lines in the layout view.

The next technique for creating uniaxial process-induced strain is the use of a tensile or compressive stressed nitride capping layer. Nitride films were among the first to be adapted for this application. By controlling the growth conditions, such as pressure, silicon nitride (SiN) layers with more than 2GPa of tensile stress and more than 2.5GPa of compressive stress have recently been developed, simultaneously improving n- and p-MOSFET performance with the so-called dual-stress liner (DSL) technique. In this approach a Si_3N_4 layer in a highly tensile stress state is uniformly deposited over the entire wafer, followed by patterning and etching the film off p-channel transistors. Thus, mechanical stress can be transferred to the channel through the silicon active area and poly-gate if a permanent stressed capping layer is deposited on a device. Since this layer serves as a stopping layer for the contact etching between the first level of metal and the transistor's S/D and gate regions, it is also known as the contact etch stop layer (CESL). CESL can contain up to 3 GPa of tensile or compressive stress, depending on the deposition conditions, thus making it an extremely effective and low-cost technique to introduce both longitudinal and out-of-plane channel stress. Similar to the e-SiGe technology, which is used to generate a compressive channel stress in p-MOSFETs, the CESL technology is a local stress technique. The stress in the MOSFET channel due to CESL arises from two sources: thermal expansion coefficient mismatch between the silicon and nitride film and intrinsic film stress caused by film shrinkage. A performance improvement of ~15% due to tensile nitride films has been reported in the literature for n-MOSFETs. The tensile stress distribution due to the nitride cap layer for n-MOSFET is shown in Figure 3.10. If one single type of capping layer (tensile or compressive) is used, one drawback of this approach is that the device of the opposite type will be degraded. To obtain better CMOS performance, two types of stressed layers should be applied to p- and n-MOSFETs accordingly. A highly compressive and tensile nitride layer is used for p- and n-MOSFETs, respectively.

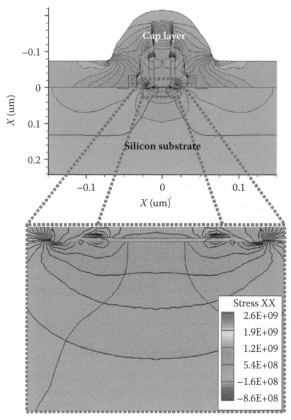

FIGURE 3.10
Simulated XX component of stress tensor (σ_{xx}) of n-MOSFETs in channel with highly tensile cap layer; distances are in micrometers. (After Maiti, T. K., Process-Induced Stress Engineering in Silicon CMOS Technology, PhD thesis, Jadavpur University, 2009.)

3.6 Silicidation

Another approach for introducing beneficial strain in p-MOSFET is by silicide-induced stress. Nickel-platinum silicide has been reported as an S/D material for strain engineering in p-MOSFETs to improve drive current performance. During the nickel-platinum silicidation process due to volume change and reaction parameters, a compressive strain can be generated in the channel region. The silicidation process incorporates a new element onto the growing layer, and hence causes stress to build up when such layers are grown under a lateral constraint. The silicide layers introduce a compressive strain on the silicon channel [9]. Ti, Ni, and Co are some of the common metals used for silicidation. Stress distribution after silicidation is shown in Figure 3.11.

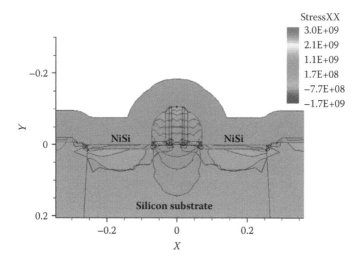

FIGURE 3.11
Stress distributions after silicidation process. (After Maiti, T. K., Process-Induced Stress Engineering in Silicon CMOS Technology, PhD thesis, Jadavpur University, 2009.)

3.7 Stress Memorisation Technique (SMT)

Another way to obtain local strain is the stress memorisation technique (SMT). The process consists of a few steps: poly-Si amorphisation, tensile CESL deposition, annealing, and liner removal. After recrystallisation, the poly-Si gate preserves some of the stress condition, even when the tensile liner is removed (Figure 3.12). This technique, although improving the performance of the n-MOSFET devices, has a detrimental effect on the p-MOSFET devices. The reason for this is not yet clearly understood.

Local strain can be applied to the channel through a stress memorisation technique. In the conventional fabrication process, the S/D silicon area and poly-gate are amorphised by S/D and extension implantation. In the SMT process, a conventional dopant activation spike anneal is performed after the deposition of a tensile capping layer, such as nitride. The stress effect is transferred from the nitride film to the channel during annealing and memorised by the recrystallisation of the S/D and poly-gate amorphised layers. Since the nitride film is disposable, a very thick capping layer can be used to increase the stress level without any process limitation. The stress effect is transferred from the nitride liner to the channel during annealing. After active area and gate recrystallisation, the stress inside the channel is memorised, even when the tensile liner is removed. The stress induced in the channel can be modulated by tuning many process parameters, such as amorphisation conditions, nitride liner thickness and intrinsic stress, annealing conditions, etc. Up to a 15% improvement in n-MOSFET drive current has been reported with the

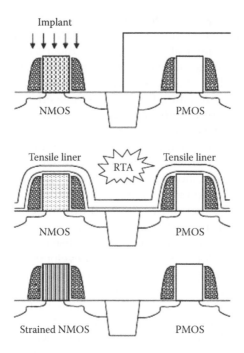

FIGURE 3.12
Simplified schematic diagrams of the main process flow steps for stress memorisation technique. An implantation of the n-MOSFET device poly-gate results in poly-Si amorphisation. Next, a tensile nitride liner is deposited and followed by rapid thermal annealing. During the annealing process, the poly-Si gate recrystallises and memorises part of the stress from the nitride liner. Next, the liner is removed. (After Shickova, A., Bias Temperature Instability Effects in Devices with Fully-Silicided Gate Stacks, Strained-Si, and Multiple-Gate Architectures, PhD thesis, Katholieke Universiteit Leuven, 2008.)

SMT technique. Further improvement by combining the SMT with the CESL technique is also possible.

3.8 Global vs. Local Strain

The global strain technique was one of the first strained Si techniques to be considered in the CMOS industry, because of its wafer-level applicability. It consists of a thin silicon layer epitaxially grown onto a $Si_{1-x}Ge_x$ strain-relaxed buffer (SRB). During its epitaxial growth, the Si layer adopts the lattice constant of the substrate material ($Si_{1-x}Ge_x$). Since the $Si_{1-x}Ge_x$ has a bigger lattice constant than the Si, the formed Si layer is under biaxial tensile strain. The amount of tensile strain in the grown Si layer depends on the Ge composition of the relaxed $Si_{1-x}Ge_x$ substrate. It is interesting to note that the hole mobility experiences an initial degradation at low Ge percentage, but then eventually at Ge > 30%, it turns to mobility improvement. This hole mobility behaviour is consistent with the classical piezoresistance theory. For mobility enhancement we consider two mechanisms: lowering of the in-plane effective mass and reduced intervalley phonon scattering. Biaxial tensile strain (less than < 1%) is not sufficient to reduce the in-plane hole effective mass. Moreover, hole intervalley scattering is not significantly reduced either, since the band splitting is less than the optical phonon energy. Thus, splitting greater than 60 meV and strain greater than 1% are necessary to suppress intervalley phonon scattering and to improve hole mobility. In order to achieve strain values suitable for CMOS applications, it is necessary to use substrates with Ge content higher than 30%. However, high Ge concentration increases the defect density in the Si layer and causes practical difficulties, such as controlling threading dislocations in practical applications.

At present, mainly two approaches are being used in obtaining the desired strain in CMOS technology. One is based on developing the strain at the substrate level before the transistor is built. This is known as the global approach, for example, strained Si on relaxed SiGe virtual substrates. The approach depends largely on materials engineering, rather than device design. Strained Si, while promising, faces several key challenges. Minimising the number of dislocations within the silicon will be important to keeping yield rates high. Maintaining the level of strain during the manufacturing process is another challenge. Also, a major drawback common to all global strain techniques for CMOS technology is that they can provide only one type of strain.

Local strained Si technology is incorporated during the transistor fabrication process via tensile/compressive capping layers or recessed epitaxial film deposition in the source/drain regions. These processes are not universal in their implementation and can be modified to a particular transistor integration scheme. The straining technique based on process is known as process-induced strain, where stress is a specified zone or local in the transistor. Process-induced uniaxial stress has advantages over biaxial stress, such as larger mobility enhancements and a smaller shift in threshold voltage. The local strain approach is found to be more promising in CMOS technology and used for high-volume production. However, the drawback of process-induced strain techniques is their strong device geometry dependence, making the scaling behaviour less predictable. The following CMOS process steps are mainly responsible for stress in the transistor channel: (1) shallow trench isolation (STI), (2) silicidation at the source/drain region, and (3) nitride contact etch stop liners (CESLs). The local strain techniques have the following advantages: (1) strain can be independently tailored to optimise performance enhancement for both n- and p-MOSFETs, (2) the threshold voltage shift is smaller in uniaxial stressed MOSFETs, (3) the stress memorisation, and (4) cheaper and more compatible with standard CMOS technology.

Even though the predominant focus of the industry in the 1990s was on biaxial stressed devices, the current focus has shifted to uniaxial stress. Starting from the 90 nm node, companies such as IBM, Intel, Texas Instruments, and Freescale have incorporated the selective epitaxial growth technique to transfer uniaxial compressive stress into the Si channel by growing a local epitaxial film of SiGe in the source and drain regions of p-MOSFETs. Depending on the proximity of the SiGe to the channel and the Ge content, 500–900 MPa stress is created in the channel. Using this technique, impressive saturation drain current enhancement up to 20–25% has been demonstrated for p-MOSFETs. A tensile Si nitride capping layer is used to introduce tensile uniaxial strain into the n-MOSFET, which enhanced the drive current by 10%.

Uniaxial strain is superior to biaxial strain in the following aspects:

1. Uniaxial stress can offer high hole mobility enhancement in both low strain and high vertical electric fields due to additive strain and confined splitting, larger two-dimensional in-plane density of states, and smaller conductivity mass.

2. Uniaxial stress-enhanced electron and hole mobilities mainly arise from reduced conductivity effective mass vs. reduced scattering for biaxial stress. Therefore, uniaxial stress provides larger drive current improvement for nanoscale short-channel devices with minimal increases in manufacturing complexity.

3. Uniaxial stress causes n-channel threshold voltage shifts that are approximately five times smaller, and thus do not require adjustment in substrate doping.

4. Process-induced uniaxial stress increases with decreasing channel length.

5. A uniaxially strained device shows much better reliability.

6. Smaller leakages arise from reduced band gap narrowing, compared with biaxial tensile stress, which causes much greater band-to-band tunneling (BTBT) leakage.

7. Significantly less strain is required for hole mobility enhancement when applying longitudinal uniaxial compression vs. in-plane biaxial tension using the conventional SiGe substrate approach. Therefore, process-induced uniaxial stress is very promising for scaling down CMOS technology per the goals of the proposed road map.

3.9 BEOL Stress: Through-Silicon Via

Stress has an impact on all of these reliability concerns. Back-end-of-the-line (BEOL) stress is very important in terms of interconnect and dielectric reliability. Going toward the 32 nm node, a key challenge is BEOL integration. In particular, dielectric reliability in the regime of low-k dielectrics is a major concern. As BEOL processing involves large amounts of stress, stress becomes a primary issue for BEOL reliability with low-k dielectrics, which are more fragile. The main BEOL reliability concerns with copper interconnects and low-k dielectrics are stress migration, time-dependent dielectric breakdown (TDDB) or bias temperature stability (BTS), delamination and crack formations, copper diffusion into low-k, and electromigration. Electromigration is mainly a result of electrical stress, but it can also get worse due to mechanical stress gradients. It has been indicated that electromigration can be impacted by mechanical stress, particularly in the interconnect extensions close to vias. Beyond mechanical stress, concerns also include thermal and electrical stress.

As continued scaling becomes increasingly difficult, 3D integration has emerged as a viable solution to achieve higher bandwidth and power efficiency. Through-silicon via (TSV), which directly connects stacked structures die to die, is one of the key techniques enabling 3D integration [11]. The advent of 3D integrated circuit (ICs) also provides the opportunity for the on-chip integration of heterogeneous devices and technologies such as memory, logic, radio frequency (RF), and sensing circuits. Multiple techniques exist to achieve 3D stacking, including wire bonding, monolithic integration, and TSVs. TSVs can be used for routing signals, power delivery, and heat extraction.

Through-silicon via is a promising and key technology to integrate chips with diverse functionalities by stacking chips vertically for implementation of 3D ICs with less space and better performance. TSVs can be used to route interdie signals, deliver power to each die, and extract heat from the dies farther away from the heat sink. Three-dimensional ICs have short interconnects among each function block, leading to better RC delay. The most important advantage of TSV structures is that the vertical interconnect successfully addresses the 2D interconnect problem by replacing long horizontal interconnects with short vertical interconnects. As a result, the RC delay, cross talk, and power dissipation will be greatly improved. However, the TSV impact on transistor performance is usually not known until the TSV process is stable and commercialised. Unlike the state-of-the-art strain technologies, such as e-SiGe or DSL, TSV thermal stress is not an intentional technique applied to improve device performance. TSV-induced stresses can lead to such effects as delamination, void formation and migration, and fracture, and can significantly affect device performance. TSV-induced substrate noise increases leakage current, which increases static power consumption. As a result, stress development is a major concern for reliability, process control, and device design.

Stress development in ICs can occur at any stage of the manufacturing process from a variety of sources. Two critical areas of Si stress development in ICs are those that can affect MOSFET performance: (1) front-end-of-the-line (FEOL) strain-engineered Si channel for increasing carrier mobility and (2) thermomechanical stress development near Cu TSVs for 3D integration. In both cases, stresses that develop in Si affect device performance; however, in the first case, these stresses are desirable, whereas in the second case, these stresses may be deleterious due to undesirable, nonuniform device performance variations. The introduction of mechanical stress in Si-based integrated circuits, whether desired or undesired, is intrinsic to IC fabrication. Stress can affect carrier mobility, either negatively or positively, depending on direction and magnitude of stress and the majority carrier type.

Although strain engineering in the channel region of a MOSFET is beneficial, in 3D architectures, the integration of Cu TSVs through the active region can induce thermomechanical residual stresses in the nearby Si, which could lead to undesirable performance variations. It is important to analyse parametric variations caused by proximity effects, such as the impact of layout on transistor stress state. Localised stress characterisation for FEOL applications commonly uses Raman spectroscopy, an all-optical technique that is applicable for measuring stress in Si based on changes in the crystalline vibrational modes. Residual stress measurements in Si are conducted using micro-Raman spectroscopy.

The thermal stress results in additional process variation and reliability issues. The management and control of this parasitic stress is important for 3D IC development. It is necessary to investigate and characterise the origins and levels of the induced stresses. The basic structure of TSV is composed of two components: conducting metals in the via and the barrier layer around metal. Copper is one of the materials frequently used to serve as an interconnect between devices due to its better immunity of electron migration (EM) and lower resistivity. The most prevalent metal used for TSVs is Cu. This choice is intuitive, considering its high conductivity, compatibility with current CMOS technology and processing, and the technological expertise developed since its introduction as the metal interconnect of choice in ICs. However, due to the large via sizes (currently 5×50 μm), there is the problem of thermally induced stresses in nearby active layer devices due to the large coefficient of thermal expansion (CTE) mismatch between Cu and Si. Thermal cycling during the manufacturing process, with temperatures reaching up to 400°C or higher, induces thermomechanical stresses in the nearby Si, which can affect mobility of nearby devices. This would lead to nonuniform device performance, which would be a function of distance from Cu TSVs.

Processes in fabricating TSVs include through-wafer via formation, deep reactive ion etching (DRIE), via filling by deposition of diffusion barrier and adhesion layers, metallisation, wafer thinning and alignment, and bonding. Manufacturing constraints associated with TSV etch and via filling processes dictate TSV size. TSV integration schemes are categorised into via-first (via formation before CMOS process), via-last (via formation after BEOL), and via-middle (via between CMOS and BEOL). During TSV fabrication, thermal stress is observed at the interface between TSV and silicon due to the mismatch in coefficients of thermal expansion between silicon substrate and metal, where copper is usually adopted as the conducting via.

Since TSV is developed based on metal-insulator-semiconductor (MIS) structure, the parasitic capacitance is different from the traditional interconnect

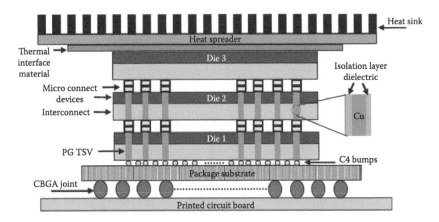

FIGURE 3.13
Three-dimensional IC, illustrating various components of a 3D system. (After Khan, N. H., Through-Silicon Via Analysis for the Design of 3-D Integrated Circuits, PhD thesis, Tufts University, 2011.)

capacitance. In addition to the influence of TSV parameters on the parasitic capacitance and the required area, one major challenge during the TSV process is the reliability due to the thermal stress [12, 13]. Tungsten, polysilicon, and copper are considered TSV conducting metals. The thermal stress originates from the mismatch in coefficients of thermal expansion between TSV fill material and silicon substrate. TSV-induced stress impact on a device and circuit performance, and its interaction with polysilicon and shallow trench isolation (STI) layout pattern density, has been studied [14, 15].

A TSV, shown in Figure 3.13, is a metal interconnect that passes through Si substrate and is electrically isolated from the substrate by a liner, an insulating material like silicon dioxide. The fabrication of TSVs may induce thermomechanical stress due to mismatch in CTEs between a TSV fill material such as copper and silicon, and most work in this area focuses on the fabrication and reliability issues.

The process steps and physical presence of TSVs, however, generate a stress-induced thermal mismatch between TSVs and the silicon bulk. The stress developed affects the performance of nearby transistors, diodes, and associated circuits. A methodology to analyse transistor characteristics and circuit performance under the influence of TSV stress is presented. The mechanical stress in the silicon is due to the mismatch in thermal expansion coefficients between the copper TSV (17.7 ppm/°C) and the surrounding silicon (3.05 ppm/°C). The large mismatch between the coefficients of thermal expansion of metallic TSV (17.5 E^{-6}/° for Cu) and Si substrate (2.5 E^{-6}/°C) results in serious reliability concerns [16, 17]. This mechanical stress can be decomposed in two directions, radial tension and tangential compression, and further affects the carrier mobility and performance of the adjacent devices through piezoresistance effects.

To fabricate a TSV of size 5μm assuming a practical aspect ratio of 10:1, the maximum die thickness will be 50 μm. A TSV is a metallic, usually copper (Cu), wire extending throughout the substrate and insulated by a dielectric material. The stress magnitude is sensitive to TSV geometry structure; when the radius of TSV metal increases, more stress is introduced and becomes saturated. This property makes it sensitive to the layout pattern. Moreover, during the process, due to the mismatch of coefficients of thermal expansion between copper and silicon, thermal stresses are observed at the interface between TSV and silicon substrate, impacting device performance of neighbouring transistors.

Stacking multiple dies to form 3D integrated circuits has emerged as a promising technology to reduce interconnects delay and power, to increase device density, and to achieve heterogeneous integration. Through-silicon vias and metallic wires that connect different dies are a key enabling technology for 3D ICs. Three-dimensional IC technology not only is capable of increased device density, but also offers heterogeneous integration of dies from disparate technologies (analogue, digital, mixed signals, sensors, antennae, and power storage) and from different technology nodes.

As TSVs create thermal stress in the substrate, stress impacts the performance of neighbouring devices. The profile of TSV-induced thermal stress in silicon follows a distribution similar to that of the leading-edge strain technology [18]. Analysis tools to quantify the impact of thermal stress on device performance and techniques to reduce this impact are required. Both TSV-induced noise and TSV-induced stress dictate the size of the keep-out zone for devices. Analyses of the two phenomena need to be performed to create new design rules for devices in 3D ICs. Thermal management is a challenge in 3D ICs. TSVs are proposed to extract heat from dies away from the heat spreader. Detailed analyses that consider dielectric liner and practical TSV placement are needed. Coupled analyses of thermal and power TSVs are required to estimate the effective substrate area dedicated to devices.

The effect of elastic anisotropy on the thermal stress distribution in Si is investigated [19]. The distribution of thermal stresses on the (001) Si wafer surface is simulated using finite element analysis (FEA). The thermal stresses on the (001) Si wafer surface are extracted from the simulation results, and the distributions of normal stresses σ_{xx} are plotted in Figure 3.14(a) and (b), with the x axis aligned with the [100] and [110] crystal directions, respectively. In the latter, the simulation is performed on the same model, except with isotropic Si, and the distribution of σ_{xx} on such an isotropic Si wafer surface is shown in Figure 3.14(c). For the sake of comparison, the stress scales in Figure 3.14(a) to (c) are normalised.

Raman characterisation of TSV-induced stress in Si has been performed [20]. Figure 3.15a displays a map of the Si Raman peak shift surrounding a 5 μm square Cu TSV. Positive Raman shifts (green, yellow, red), which represent compressive stresses, are observed within ~2 μm of the Cu TSV. Negative Raman shifts (blue), representing tensile stresses, are observed

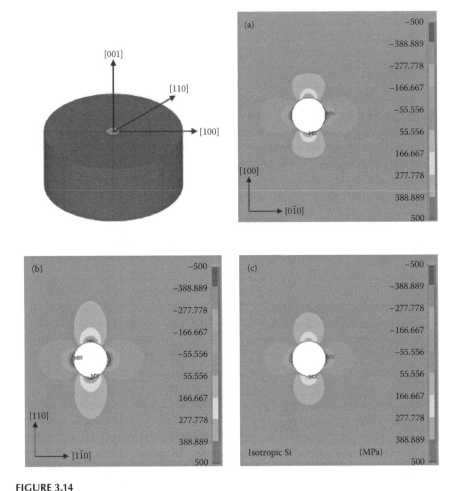

FIGURE 3.14
Thermal stress distribution on the Si wafer surface: (a) σ_{xx} with x axis along [100] crystal direction, (b) σ_{xx} with x axis along [110] crystal direction, and (c) σ_{xx} on an isotropic Si wafer ($D_f = 20$ μm, $\Delta T = -200°C$). (After Lu, K. H., Thermo-Mechanical Reliability of 3-D Interconnects Containing Through-Silicon-Vias (TSVs), PhD thesis, University of Texas at Austin, 2010.)

at greater distances (>2 μm) from the TSV faces. A fourfold symmetrical Si Raman peak distribution around the TSV mirrors the TSV shape, but may also be influenced by the cubic anisotropy in the Si elastic stiffness matrix.

Figure 3.16 displays a Si Raman peak shift map and a corresponding linear biaxial stress map from a 5 μm diameter, round Cu TSV (wafer A). A region of compressive stress is observed in the Si within ~1–2 μm of the Cu TSV. In regions farther away (>2 μm), the fourfold symmetric tensile stress distribution is observed. The presence of the fourfold symmetric stress field here cannot be attributed to the TSV geometry. Consequently, it arises solely from

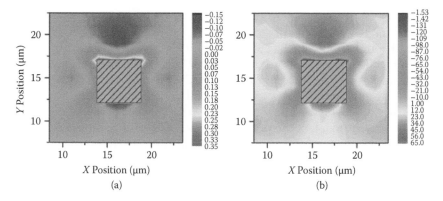

FIGURE 3.15

(a) Si Raman peak shift map surrounding a 5 ´ 5 µm square TSV. (b) Linear conversion of Si Raman peak shift to biaxial stress displays a fourfold symmetrical biaxial stress distribution with high compressive stresses concentrated near the corners of the TSV. (After McDonough, C. J., Applications of Raman Spectroscopy for Silicon Stress Characterization in Integrated Circuits, PhD thesis, University at Albany, State University of New York, 2011.)

the Si anisotropy. Tensile stress relaxation occurs at larger distances (>15 µm) from the TSV in the <110> directions. The diminished compressive stress region in the <100> directions is consistent with the orientation dependence of Si Young's modulus. Due to high stress concentrations at the corners of square TSVs, a round TSV geometry is better suited to minimise the stress

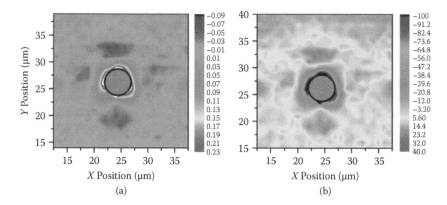

FIGURE 3.16

(a) Si Raman peak shift map surrounding an isolated, 5 × 25 µm Cu TSV. (b) Linear conversion of Si Raman peak displays a uniform distribution of stress. Some compression of the Si is measured near the empty TSV. (After McDonough, C. J., Applications of Raman Spectroscopy for Silicon Stress Characterization in Integrated Circuits, PhD thesis, University at Albany, State University of New York, 2011.)

field in nearby Si. Due to these and other reasons a round TSV geometry was adopted for TSV test structures.

The through-silicon via (TSV) proximity effect on transistor performance has been reported by Yang et al. [21]. The authors have evaluated the electrical performance on a 130 nm CMOS platform. In their work, stacked circuits with TSVs were fabricated on 130 nm CMOS technology platform. MOSFETs, with TSV proximity in several patterns, were electrically evaluated to detect the possible impact. The TSVs are placed close to the channel regions of the MOSFETs. The distance between the edge of a TSV and the edge of a transistor channel was set as 1.1 μm to avoid damaging the device. Designs with multiple TSVs close to a transistor have been investigated as well.

3.10 TSV Modelling

Modelling of TSV is currently an active research area. Thermomechanical reliability in 3D interconnects containing TSVs is of serious concern and includes (1) a thermal stress measurement, (2) TSV-induced thermal stresses in a Si matrix and the impact on electrical performance of devices, and (3) TSV-induced thermal stresses at the TSV/Si interface. TSV-induced stresses in Si are calculated combining analytical solutions and FEA simulations. A 3D semianalytical stress model valid for high-aspect-ratio TSVs has also been developed [22] to characterise the near-surface stress distribution. The reliability issues, such as carrier mobility change in transistors, the interfacial delamination of TSVs, and thermal stress interactions between TSVs induced by the thermal stresses, have been studied by several research groups [23]. A stress model has been developed to model TSV-induced stress effect to predict the influence of the stress and help designers optimise the circuit performance. Stress levels calculated via finite element analysis have shown stress levels can reach the order of several hundred MPa. In this regime, the fractional change in carrier mobility, $\Delta\mu/\mu$, can be found by the piezoresistance constants of Si. A 100 MPa stress can induce up to 7% mobility change in Si. Therefore, it will be important to understand the thermally induced stress development throughout the process flow to determine keep-out zones that will determine minimum distances between TSVs and nearby MOSFETs for maintaining acceptable mobility deviations. Interfacial delamination of TSVs was found to be mainly driven by a shear stress concentration at the TSV/Si interface. Change in mobility due to TSV stress in transistors was found to be sensitive to the normal stresses near the Si wafer surface. The surface area of a high-mobility change is defined as the keep-out zone (KOZ) for transistors. KOZ is mainly controlled by the TSV geometry and the materials used. FEA simulations are carried out to calculate the area of KOZ surrounding TSVs. The area of KOZ has been found to be mainly determined by the channel direction of the transistor as a result of anisotropic piezoresistivity effects. Both finite element analysis and analytical models have been proposed to characterise stress induced by TSVs. The analytical 2D radial stress model was employed to address the TSV thermomechanical stress effect on device performance [19]. Tanaka et al. [24] reported that MOS transistor operation after both the postprocessing of TSVs and postassembly was slightly affected by mechanical stress depending on the distance from a TSV to a MOS transistor.

Mercha et al. [25] have experimentally demonstrated a significant impact of TSVs on the adjacent transistors, with up to 30% I_{dsat} shift due to TSV stress. A FEM model has been proposed using the measured TSV Cu properties (CTE, stress in Cu as a function of temperature, plastic behaviour, etc.) and relevant information, such as processing temperature profiles, and was used to predict the mechanical stress tensor throughout the silicon die. Excellent modelling accuracy has been achieved within 0.5% of the measured I_{dsat} values. The keep-out zones (KOZs) for a large matrix of TSVs are over 200 µm for analogue circuits and 20 µm for digital circuits. It has been shown that the complex interaction of stress components makes it difficult to use simple design rules without sacrificing large layout area. Numerical 3D stress analysis can be used to accurately estimate KOZ for different TSV placements.

Low capacitance and resistance and high-density integration are the main desired features in a TSV structure. Both the required area and interconnect performance heavily rely on TSV process and structures, including the design of the metal radius, the thickness of the barrier layer, and the doping concentration in the silicon substrate. Figure 3.17(a) shows the impact of copper radius on threshold voltage, resistance, and the smallest capacitance. As the radius of copper (r_{TSV}) increases, the resistance decreases because of the larger copper cross section, but the capacitance increases. Moreover, the threshold voltage drops because capacitance increases with the radius of copper. The trade-offs between the RC delay and the area requirement are shown for various copper radiuses in Figure 3.17(b). RC product decreases as a larger copper radius is introduced, implying the impact of resistance reduction is stronger than the increasing capacitance. On the other hand, with the growth of copper radius, a larger TSV area is demanded, showing a trade-off between area and TSV performance. Figure 3.18 shows the threshold voltage changes with the radius of TSV metal. As the radius increases, the threshold voltage decreases; as oxide thickness increases, the threshold voltage increases. For a high-frequency operation, the threshold voltage should be smaller than the applied voltage, as shown in the highlighted area in Figure 3.18.

Although the TSV stress is not intentionally applied to impact the device performance, there is a keep-out zone (KOZ) to keep devices unaffected by

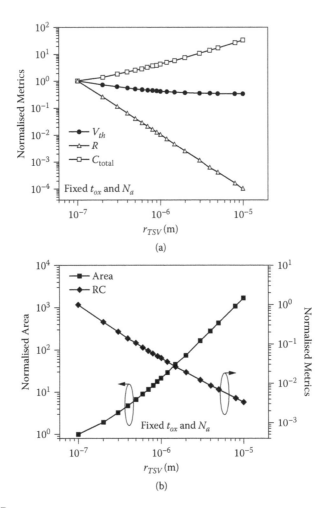

FIGURE 3.17
(a) Threshold voltage, resistance, and capacitance change with various metal radii. (b) Trade-off between RC delay and TSV area. (After Wang, C.-C., Predictive Modelling for Extremely Scaled CMOS and Post Silicon Devices, PhD thesis, Arizona State University, 2011.)

TSV stress. The keep-out zone can be defined in the region from peak stress to the bottom stress. Within the keep-out zone, the device performance varies from location to location, while device performance is not sensitive to the stress effect outside the keep-out zone. Inside the keep-out zone, the stress is significant to impact the device performance, so there are no devices allowed in this zone, leading to more area cost during the TSV process. However, KOZ can be utilised with a stress-aware design for area efficiency. Figure 3.19(a) shows the mobility variation changes with the distance. The mobility enhancement factor decays over the distance from the TSV edge; the farther the device is located, the less stress effect and less mobility variation.

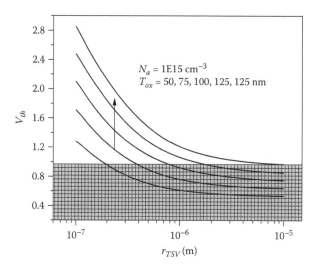

FIGURE 3.18
Threshold voltage varies with the radius of copper for different oxide thickness. (After Wang, C.-C., Predictive Modelling for Extremely Scaled CMOS and Post Silicon Devices, PhD thesis, Arizona State University, 2011.)

Moreover, as the TSV radius increases, the stress effect becomes stronger and finally saturates. On the other hand, when the device is located outside the keep-out zone, the device performance is very stable and hardly affected by the stress. Figure 3.19(b) shows the mobility variation factor changes with TSV radius. As the radius increases, the bottom stress grows up toward saturation. This phenomenon is similar to that the stress is saturated in e-SiGe

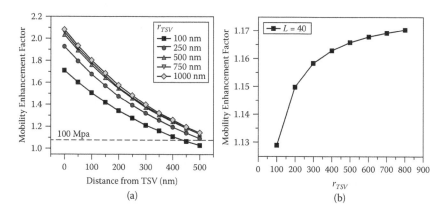

FIGURE 3.19
Characteristics for a device placed (a) inside the keep-out zone and (b) outside the keep-out zone. (After Wang, C.-C., Predictive Modelling for Extremely Scaled CMOS and Post Silicon Devices, PhD thesis, Arizona State University, 2011.)

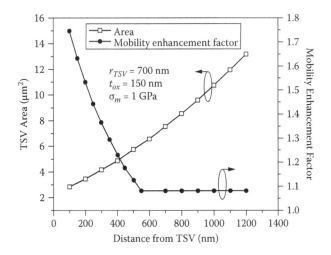

FIGURE 3.20
Trade-off between TSV area and the influence of mobility enhancement factor. (After Wang, C.-C., Predictive Modelling for Extremely Scaled CMOS and Post Silicon Devices, PhD thesis, Arizona State University, 2011.)

technology when source/drain length increases, and the same modelling approach is applicable to the TSV-induced thermal stress effect.

With the assistance of the stress model, the impact of TSV thermal stress on mobility variation may be predicted. Figure 3.20 shows the mobility variation with the distance from the TSV edge. The mobility varies significantly inside the keep-out zone, while the stays stable out of the keep-out zone. To keep devices unaffected from the thermal stress, the area of the keep-out zone is required. More KOZ area reduces the impacts of thermal stress on devices, ensuring the stable process variation. However, the keep-out zone can be utilised with a stress-aware design approach if the mobility variation can be well modeled, illustrating the opportunities from a joint device-design perspective.

For an arbitrary criterion for KOZ, assuming equivalent to 10% change in mobility, the area of KOZ surrounding the TSV has been calculated by Lu [19]. The effects of TSV diameter D_f and wafer thickness on KOZ are shown in Figures 3.21(a) and (b), respectively. It is seen that the diameter has a significant effect on the area of KOZ for p-MOSFET, increasing approximately with the square of D_f. Figure 3.21(b) shows that the area of KOZ for p-MOSFET initially increases with the wafer thickness H, and then reaches a stable value if the wafer thickness is greater than $5D_f$.

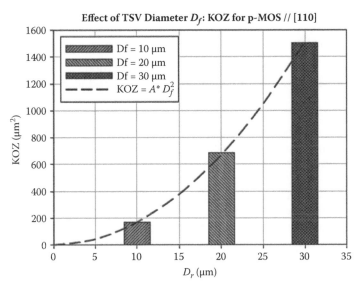

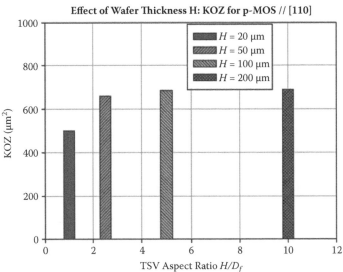

FIGURE 3.21
(a) Effect of TSV diameter on KOZ for p-MOSFETs along [110] crystal direction. (b) Effect of wafer thickness on KOZ for p-MOSFETs along [110] crystal direction ($D_f = 20$ μm, $\Delta T = -180°C$). (After Lu, K. H., Thermo-Mechanical Reliability of 3-D Interconnects Containing Through-Silicon-Vias (TSVs), PhD thesis, University of Texas at Austin, 2010.)

3.11 Summary

A brief and comprehensive review on the methods used to introduce strain in CMOS transistors, such as process-introduced uniaxial strain, is presented. The semiconductor industry has adopted uniaxial strain over biaxial strain because of higher performance improvement. Starting with the 90 nm technology node, uniaxial stress has successfully been integrated into the mainstream MOSFET process flow. Encouraged by the strain-enhanced planar MOSFETs, now application of uniaxial stress to multigate devices is being contemplated with metal gate and high-k dielectric as a performance booster. However, the drawbacks of uniaxial stress, e.g., the localised stress dependence on device size and defects from additional processes, may affect the overall performance and must be addressed carefully. Some new techniques to introduce stress, such as the stress memorisation technique and the stress proximity technique, are also discussed. Uniaxial stress is found to have several advantages over biaxial stress. Key challenge for integrating stress in CMOS manufacturing technologies is discussed.

Review Questions

1. What is process-induced strain?
2. Discuss the influence of Ge content on stress developed in the channel region.
3. What are the traditional scaling limiters and their implications on nanoscale MOSFETs?
4. What is the purpose of embedded SiGe in the source/drain region of a p-MOSFET?
5. Discuss the influence of Ge content on stress developed in the channel region.
6. What is role of graded Ge content in the buffer layer for producing strained Si films?
7. Compare the local and global strains.
8. What is hybrid orientation technology?
9. How is uniaxial process-induced strain used to enhance mobility?
10. Why is low field mobility important for nanoscale transistors?

References

1. C. K. Maiti, S. Chattopadhyay, and L. K. Bera, *Strained-Si Heterostructure Field-Effect Devices*, CRC Press (Taylor & Francis), Boca Raton, FL, 2007.

2. S. E. Thompson, M. Armstrong, C. Auth, S. Cea, R. Chau, G. Glass, T. Hoffman, J. Klaus, Z. Ma, M. Bohr, and Y. El-Mansy. A 90nm Logic Technology Featuring Strained-Silicon, *IEEE Trans. Electron Dev.*, 51, 1790–1797, 2004.

3. S. E. Thompson, G. Sun, Y. Choi, and T. Nishida, Uniaxial-Process-Induced Strained-Si: Extending the CMOS Roadmap, *IEEE Trans. Electron. Dev.*, 53, 1010–1020, 2006.

4. H. H. Hall, J. Bardeen, and G. L. Pearson, The Effects of Pressure and Temperature on the Resistance of Junctions in Germanium, *Phys. Rev.*, 84, 129–132, 1951.

5. C. S. Smith, Piezoresistance Effect in Germanium and Silicon, *Phys. Rev.*, 94, 42–49, 1954.

6. J. Welser, J. L. Hoyt, and J. F. Gibbons, NMOS and PMOS Transistors Fabricated in Strained Silicon/Relaxed Silicon-Germanium Structures, *IEEE IEDM Tech. Dig.*, 1000–1002, 1992.

7. J. Welser, J. L. Hoyt, and J. F. Gibbons, Electron Mobility Enhancement in Strained-Si N-Type Metal-Oxide-Semiconductor Field-Effect Transistors, *IEEE Electron Device Lett.*, 15, 100–102, 1994.

8. P. Verheyen, N. Collaert, R. Rooyackers, R. Loo, D. Shamiryan, A. De Keersgieter, G. Eneman, F. Leys, A. Dixit, M. Goodwin, Y. S. Yim, M. Caymax, K. De Meyer, P. Absil, M. Jurczak, and S. Biesemans, 25% Drive Current Improvement for p-Type Multiple Gate FET (MuGFET) Devices by the Introduction of Recessed $Si_{0.8}Ge_{0.2}$ in the Source and Drain Regions, *Proc. Symp. VLSI Technol.*, 194–195, 2005.

9. P. R. Chidambaram, C. Bowen, S. Chakravarthi, C. Machala, and R. Wise, Fundamentals of Silicon Material Properties for Successful Exploitation of Strain Engineering in Modern CMOS Manufacturing, *IEEE Trans. Electron Dev.*, 53, 944–964, 2006.

10. H. Tsuno, K. Anzai, M. Matsumura, S. Minami, A. Honjo, H. Koike, Y. Hiura, A. Takeo, W. Fu, Y. Fukuzaki, M. Kanno, H. Ansai, and N. Nagashima, Advanced Analysis and Modelling of MOSFET Characteristic Fluctuation Caused by Layout Variation, *Proc. Symp. VLSI Technol.*, 204–205, 2007.

11. K. Takahashi and M. Sekiguchi, Through Silicon Via and 3-D Wafer/Chip Stacking Technology, *Proc. Symp. VLSI Circuits*, 89–92, 2006.

12. K. H. Lu, X. Zhang, S.-K. Ryu, J. Im, R. Huang, and P. S. Ho, Thermo-Mechanical Reliability of 3-D ICs Containing Through Silicon Vias, *Proc. IEEE Electronic Components Technol. Conf.*, 630–634, 2009.

13. C. S. Selvanayagam, X. Zhang, R. Rajoo, and D. Pinjala, Modelling Stress in Silicon with TSVs and Its Effect on Mobility, *Proc. IEEE Electronics Packaging Technol. Conf.*, 612–618, 2009.

14. N. H. Khan, Through-Silicon Via Analysis for the Design of 3-D Integrated Circuits, PhD thesis, Tufts University, 2011.

15. L. Yu, A Study of Through-Silicon-Via (TSV) Induced Transistor Variation, Master of Science thesis, Massachusetts Institute of Technology, 2011.

16. C. S. Selvanayagam, J. H. Lau, X. Zhang, S. Seah, K. Vaidyanathan, and T. C. Chai, Nonlinear Thermal Stress/Strain Analyses of Copper Filled TSV (Through Silicon Via) and Their Flip-Chip Microbumps, *IEEE Trans. Adv. Packaging*, 32, 720–728, 2009.
17. B. Wunderle, R. Mrossko, O.Wittler, E. Kaulfersch, P. Ramm, B. Michel, and H. Reichl, Thermo-Mechanical Reliability of 3-D-Integrated Microstructures in Stacked Silicon, *Proc. Mater. Res. Soc. Symp.*, 67, 970–974, 2007.
18. T. Chidambaram, C. McDonough, R. Geer, and W. Wang, TSV Stress Testing and Modelling for 3D IC Applications, *Proc. Physical Failure Anal. Integrated Circuits*, 727–730, 2009.
19. K. H. Lu, Thermo-Mechanical Reliability of 3-D Interconnects Containing Through-Silicon-Vias (TSVs), PhD thesis, University of Texas at Austin, 2010.
20. C. J. McDonough, Applications of Raman Spectroscopy for Silicon Stress Characterization in Integrated Circuits, PhD thesis, University at Albany, State University of New York, 2011.
21. Y. Yang, G. Katti, R. Labie, Y. Travaly, B. Verlinden, and I. De Wolf, Electrical Evaluation of 130-nm MOSFETs with TSV Proximity in 3D-SIC Structure, *Proc. Interconnect Technology Conf. (IITC)*, 1–3, 2010.
22. S. Ryu, K. Lu, X. Zhang, J. Im, P. Ho, and R. Huang, Impact of Near-Surface Thermal Stresses on Interfacial Reliability of Through-Silicon-Vias for 3-D Interconnects, *IEEE Trans. Dev. Mater. Reliab.*, 11, 35–43, 2011.
23. J. H. Lau, Evolution and Outlook of TSV and 3D IC/Si Integration, *Proc. 12th Electronics Packaging Technology Conf.*, 560–570, 2010.
24. N. Tanaka, M. Kawashita, Y. Yoshimura, T. Uematsu, M. Fujisawa, H. Shimokawa, N. Kinoshita, T. Naito, T. Kikuchi, and T. Akazawa, Characterization of MOS Transistors after TSV Fabrication and 3D-Assembly, *Proc. 2nd Electronics Syst. Integration*, 131–134, 2008.
25. A. Mercha, G. Van der Plas, V. Moroz, I. De Wolf, P. Asimakopoulos, N. Minas, S. Domae, D. Perry, M. Choi, A. Redolfi, C. Okoro, Y. Yang, J. Van Olmen, S. Thangaraju, D. Tezcan, P. Soussan, J. Cho, A. Yakovlev, P. Marchal, Y. Travaly, E. Beyne, S. Biesemans, and B. Swinnen, Comprehensive Analysis of the Impact of Single and Arrays of Through Silicon Vias Induced Stress on High-k/Metal Gate CMOS Performance, *IEEE IEDM Tech. Dig.*, 2.2.1–2.2.4, 2010.

4

Electronic Properties of Strain-Engineered Semiconductors

It wasn't until the early 1980s when engineers and scientists started to realise that strain could be a powerful tool to modify the band structure of semiconductors. The band structure determines several important characteristics, in particular, its electronic and optical properties. The deformation potential theory, which defines the concept of strain-induced energy shift of the semiconductor, was first developed to account for the coupling between the acoustic waves and electrons in solids by Bardeen and Shockley [1]. It has been stated that the local shift of energy bands by the acoustic phonon would be produced by an equivalent extrinsic strain; hence, the energy shifts by both intrinsic and extrinsic strain can be described in the same deformation potential framework. Piezoresistance coefficients are widely used due to their simplicity in representing the semiconductor transport properties under strain. The first experimental work that reported strain effects on semiconductor transport was by Smith [2], who measured the piezoresistance coefficients for n- and p-type strained bulk silicon and germanium in 1954.

Strained Si technologies have been widely studied as a new promising scaling vector (mobility scaling) to improve on-state drive current without degrading off-state leakage current. Mobilities of both electrons and holes can be improved by applying stress to induce appropriate strain in the channel, e.g., tensile strain for n-MOSFETs and compressive strain for p-MOSFETs. In this chapter, the physics of strained Si is reviewed using electronic band structures, and the simple piezoresistive (PR) model is also introduced to quantify mobility enhancement induced by strain. Uniaxial or biaxial tensile strain changes the electronic band structure of Si, leading to carrier repopulation and band splitting between subvalleys, resulting in a change in effective carrier mobility. Strain enhances the carrier mobility, which is given by $\mu = q\tau/m^*$, by reducing the conductivity effective mass (m^*) or increasing the relaxation time (τ). The biaxial tensile strain also improves hole mobility by reducing hole conductivity effective mass and suppressing intervalley scattering. With strain, the hole conductivity effective mass becomes anisotropic due to band warping, and holes preferentially occupy higher energy light-hole (LH) valleys due to energy splitting. The hole mobility enhancement under biaxial tensile strain is mainly due to the large reduction of hole intervalley scattering from energy splitting between light-hole (LH) and heavy-hole (HH) bands, especially for stress higher than 1 GPa. However, this energy splitting between LH and HH bands decreases at high vertical electric field (E_{eff}) due to the quantum mechanical confinement effect.

Uniaxial strain along the silicon channel has been also widely used to enhance both electron and hole mobilities. Similar to biaxial tensile strain, uniaxial tensile strain improves electron mobility by reducing the net in-plane conductivity effective mass by band repopulation; i.e., electrons preferentially occupy the four lower energy valleys (unstrained valleys) with small in-plane effective mass. Intervalley scattering is also suppressed by the energy splitting between strained valleys (two in-plane valleys) and unstrained valleys (two in-plane and two out-of-plane valleys), but this is smaller than that for biaxial strain, giving an advantageously small n-MOSFET V_{th} shift. The reduction of hole intervalley scattering from energy splitting between light-hole (LH) and heavy-hole (HH) bands also improves hole mobility. In particular, the band splitting is maintained even at high E_{eff} due to anisotropic out-of-plane hole effective masses of top (LH) and second (HH) bands. On the other hand, uniaxial compressive strain improves hole mobility more effectively than biaxial tensile strain. Under uniaxial compressive strain, the net in-plane hole conductivity effective mass becomes much smaller due to reduced in-plane hole conductivity effective mass of the top (LH) band, while biaxial tensile strain shows the opposite top band curvature.

Uniaxial tensile strain also enhances electron mobility by reducing effective mass and suppressing intervalley scattering. However, the energy splitting between strained valleys and unstrained valleys is smaller than that of biaxial strain. In this chapter, the basic physical definitions, such as the strain and stress tensors, are introduced and it is shown how they are related. Different methods of calculating the effect of strain on the band structure are presented. Since carrier mobility is a key parameter for the simulation of the electrical characteristics of semiconductor devices, several analytical models have been developed capturing the dependence of mobility on temperature, doping, and electric field [3–7]. All these models are developed for unstrained Si. For device simulation of strained-Si MOSFETs, different types of strain-related mobility models need to be developed [8, 9]. A simple piezoresistance model is introduced to quantify strain-induced mobility enhancement. The piezoresistance coefficients with arbitrary crystallographic orientations can be obtained by an appropriate coordinate transformation.

4.1 Basics of Stress Engineering

Since the primary focus is on strain-engineered MOSFETs, it is essential to understand the basics of engineering mechanics like stress, strain, and mechanical properties of the semiconductor involved. Within the elastic limit, the property of solid materials to deform under the application of an external force and to regain their original shape after the force is removed is referred to as elasticity. It is Hooke's law, which describes the elastic relationship between the mechanical constraint and deformation that the material will undergo. The external force applied on a specified area is known as stress, while the amount of deformation is called the strain. In the following, the theory of stress, strain, and their interdependence is briefly discussed.

4.1.1 Stress

The stress (σ) at a point may be determined by considering a small element of the body enclosed by area (ΔA) on which forces act (ΔP), and its unit is Pascal (Pa). By making the element infinitesimally small, the stress (σ) vector is defined as the limit

$$\sigma = \lim_{\Delta A \to 0} \frac{\Delta F}{\Delta A} = \frac{dF}{dA} \tag{4.1}$$

From Figure 4.1 one can observe that the force acting on a plane can be decomposed into a force within the plane, the shear components, and one perpendicular to the plane, the normal component.

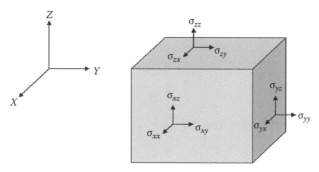

FIGURE 4.1
Stress components acting on an infinitesimal cube.

The shear stress can be further decomposed into two orthogonal force components giving rise to three total stress components acting on each plane. Figure 4.1 shows the normal and shear stresses in X, Y, and Z directions acting on different planes of the cube. The first subscript identifies the face on which the stress is acting, and the second subscript identifies the direction. The σ_{ij} components are the normal stresses, while the σ_{ij} components are the shear stresses.

4.2 Stress–Strain Relationships

Generally, the stress tensor σ is symmetric 3×3 matrices. Therefore, it only has six independent components. With the index transformation rule, it can be written in a six-component vector notation, $\sigma_{11} \rightarrow \sigma_1$, $\sigma_{22} \rightarrow \sigma_2$, $\sigma_{33} \rightarrow \sigma_3$, $\sigma_{23} \rightarrow (\sigma_4)/2$, $\sigma_{13} \rightarrow (\sigma_5)/2$, $\sigma_{12} \rightarrow (\sigma_6)/2$, that simplifies tensor expressions. For example, to compute the strain tensor (which is needed for the deformation potential model development), the generalised Hooke's law for anisotropic materials is applied as

$$\varepsilon_{ij} = \sum_{j=1}^{6} S_{ij}\sigma_j \qquad (4.2)$$

where S_{ij} is the elasticity modulus. In crystals with cubic symmetry such as silicon, the number of independent coefficients of the elasticity tensor (as other material property tensors) reduces to three by rotating the coordinate system parallel to the high-symmetric axes of the crystal [10]. This gives the following elasticity tensor $\bar{S}$ as

$$\bar{S} = \begin{bmatrix} S_{11} & S_{12} & S_{12} & 0 & 0 & 0 \\ S_{12} & S_{11} & S_{12} & 0 & 0 & 0 \\ S_{12} & S_{12} & S_{11} & 0 & 0 & 0 \\ 0 & 0 & 0 & S_{44} & 0 & 0 \\ 0 & 0 & 0 & 0 & S_{44} & 0 \\ 0 & 0 & 0 & 0 & 0 & S_{44} \end{bmatrix} \qquad (4.3)$$

where the coefficients S_{11}, S_{12}, and S_{44} correspond to parallel, perpendicular, and shear components, respectively. In Sentaurus Device, the stress tensor has been defined in the stress coordinate system $(\vec{e}_1, \vec{e}_2, \vec{e}_3)$ [8]. To transfer this tensor to another coordinate system (for example, the crystal system

$(\vec{e}_1', \vec{e}_2', \vec{e}_3')$, which is a common operation), the following transformation rule between two coordinate systems is applied:

$$\sigma_{12}' = a_{ik} a_{jl} \sigma_{kl} \tag{4.4}$$

where a is the rotation matrix

$$a_{ik} = (\vec{e}_i', \vec{e}_k) / (\|\vec{e}_i'\| \|\vec{e}_k\|) \tag{4.5}$$

4.2.1 Modelling of Stress Generation

In this section, a software-based approach is presented to engineer the stress generated in the MOSFET channel, with an ultimate goal of enhancing the device performance. This is the common and most effective approach adopted in the semiconductor industry to model process-induced stress. SProcess (Sentaurus Process) is a process simulator [11] that simulates standard process simulation steps like oxidation, diffusion, implantation, etching, etc. SProcess accepts sequence of commands at the command prompt. The process flow is simulated by issuing a sequence of commands that correspond to the individual process steps. SProcess is written in Tool Command Language (Tcl), so all Tcl commands and functionalities are supported by the software. SProcess supports several mechanical models to compute mechanical stress, such as viscous, visco-elastic, elastic, plasticity, etc. All simulations in this work are performed using the elastic model.

4.3 Strain-Engineered MOSFETs: Current

The switching speed of an ideal transistor can be increased primarily by two ways: physical gate length scaling and carrier mobility enhancement [12]. Strained Si increases the switching speed solely by enhancing the carrier mobility. The carrier mobility is given by [13]

$$\mu = q \frac{\tau}{m^*} \tag{4.6}$$

where $1/\tau$ = scattering rate and m^* = conductivity effective mass. The carrier mobility is enhanced by strain by reducing the effective mass or the scattering rate. Electron mobility is enhanced by both phenomena, while for holes, only mass change due to band warping is known to play a significant role at the current stress levels in production. The simple drain current (I_D)

expressions for long-channel MOSFETs operating in linear and saturation regions are given by

$$I_{D(lin)} = \mu.C_{OX} \frac{W}{L} \left((V_{GS} - V_{TH})V_{DS} - \frac{1}{2}V_{DS}^2 \right)$$

$$I_{D(sat)} = \frac{\mu.C_{OX}}{2} \frac{W}{L} (V_{GS} - V_{TH})^2$$

(4.7)

In the case of the nanometer region, the carrier transport in the device becomes [14]

$$I_D = C_{ox}W(V_{GS} - V_{TH}) \times v_{thermal} \frac{1 - R_C}{1 + R_C} \left[\frac{1 - e^{-\frac{qV_{DS}}{kT}}}{1 + \frac{1 - R_C}{1 + R_C} e^{-\frac{qV_{DS}}{kT}}} \right]$$

(4.8)

where R_C is the backscattering coefficient (a real number lying between 0 and 1), which is linked to the degree of transport ballisticity, and $V_{thermal}$ is the thermal velocity of the carriers and is given by [15]

$$v_{thermal} = \sqrt{\frac{2kT}{\pi m^*}} = \sqrt{\frac{2kT}{\pi} \left(\frac{P_x}{m_x} + \frac{P_y}{m_y} + \frac{P_z}{m_z} \right)}$$

(4.9)

Probability factors (P_x, P_y, and P_z) can be calculated using the following equation:

$$P_i = \frac{e^{-\Delta E_{\alpha,i}/kT}}{\sum_{i=x,y,z} e^{-\Delta E_{\alpha,i}/kT}} \qquad \{\alpha = C, V\}$$

(4.10)

Using a strained-induced band structure from Equations (4.16) and (4.17), the intrinsic carrier concentration is calculated by averaging contributions of different bands as

$$n_i^2 = \sum_{\substack{i=x,y,z \\ j=hh,lh}} N_{C,i}N_{V,j} \exp\left[-\frac{(E_g + \Delta E_{C,i} - \Delta E_{V,j})}{kT} \right]$$

(4.11)

The carrier distribution for strained Si is used to calculate the probabilities used in Equation (4.9) to have an electron in the *i*th states.

4.4 Energy Gap and Band Structure

4.4.1 Bulk Si Band Structure

The structure of crystalline silicon is a network face-centred cubic (FCC), with a diamond-like structure, and is illustrated in Figure 4.2.

Each node in the network is composed of two atoms placed in positions (0, 0, 0) and (1/4, 1/4, 1/4). The basic cell (cell Wigner–Seitz) reciprocal lattice, commonly known as the Brillouin zone, is represented in Figure 4.3 and depends on the wave vector K.

An electron in a solid is defined by its energy E and its wave function ψ linked by the Schrödinger equation (4.13) as

$$H\Psi = E\Psi \tag{4.12}$$

where H is the Hamiltonian of the system. In a periodic crystal lattice, the structure of the band is described in reciprocal space by the relations of dispersion $E(K)$. There are several methods for calculating the effect of strain on $E(K)$:

1. Ab initio *method*: Based on the approximation of the local density (LDA) and the density functional theory (DFT) within the framework of LDA.

2. *Deformation potential theory*: Developed by Bardeen and Shockley. The perturbation caused by strain is attributed to an additional Hamiltonian that is linearly proportional to the deformation potential operator and strain. First-order perturbation theory is used to calculate the effect of strain on the band structure, and analytical expressions for the strain-induced energy shifts of the conduction and valence bands can be obtained.

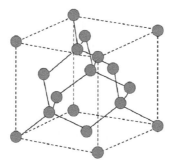

FIGURE 4.2
Crystalline structure of silicon.

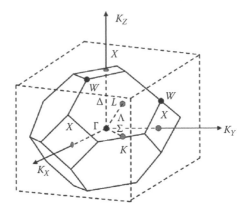

FIGURE 4.3
Brillouin zone of silicon. The points *W, L, K,* and *X* represent the principal directions.

3. *k.p method*: The main feature of the k.p method is to capture the deformation of the shape of the energy bands under strain [16].
4. *Empirical pseudopotential method* (EPM): Includes nonlocal effects. Spin-orbit coupling is frequently used to calculate the band structure of semiconductors [17–20].

4.5 Silicon Conduction Band

The minimum of the band of conduction is on the way Γ-X, which corresponds to direction <100>. Silicon being a cubic crystal, the directions <100>, <010>, <001>, <100>, <010>, and <001> are equivalent and give us six equivalent minima, also called valleys, Δ. Figure 4.4 shows isoenergy surfaces around each of the six minimum conduction valleys. Six ellipsoidal surfaces are arranged according to the six directions equivalent to <100>.

The wave functions are solutions of plane waves reflecting the decentralised nature of the particles. Relation dispersions are parabolic and written as

$$E(k) = \frac{\hbar^2 k^2}{2m} \tag{4.13}$$

By breaking up this equation along the three axes, we obtain a relation of the type of a general equation of an ellipsoid:

$$E(k) = \frac{\hbar^2}{2m_0}\left(\frac{k_x^2}{m_x} + \frac{k_y^2}{m_y} + \frac{k_z^2}{m_z}\right) \tag{4.14}$$

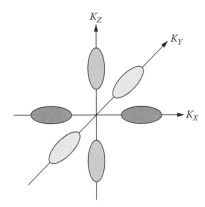

FIGURE 4.4
Ellipsoids of mass valleys Δ along crystallographic principal directions.

with m_x, m_y, and m_z being the effective masses, according to the wave vectors k_x, k_y, and k_z, respectively. This equation describes in the space of k with constant energy an ellipsoid of mass.

4.6 Silicon Valence Band

The valence band filled up with holes; its maximum is centred in Γ point. Figure 4.5 shows the detail of the structure of bands in the directions <110> and <100>. Three bands coexist of (1) heavy holes (HHs), (2) light holes (LHs), and (3) spin-orbit (SO) holes.

We notice that the bands HH and LH are degenerated into their maximum but do not have the same ray of curve, and for the spin-orbit band are one; it is located at the lower part of the two others. Contrary to the bands of conduction, the valence bands are strongly anisotropic (Figure 4.6). In particular, heavy holes that occupy most of the valence band have higher mass. Thus, the choice of the direction of transport becomes important for p-MOSFETs.

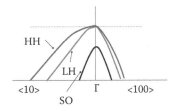

FIGURE 4.5
Valence bands in the directions <100> and <110>.

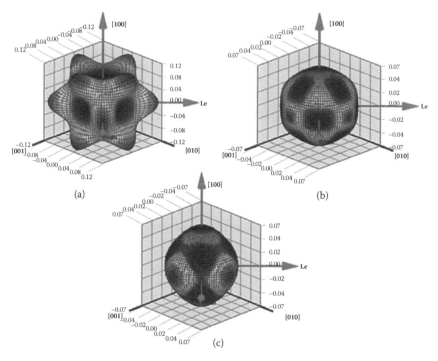

FIGURE 4.6

Isoenergy surfaces around the maximum of the valence band represented in reciprocal space for silicon: (a) heavy holes, (b) light holes, and (c) spin-orbit holes.

The isoenergy surfaces are calculated using the k.p model [21–24]. For simulation, however, we used a very simplified version implemented in MASTAR [15, 25].

4.7 Band Structure under Stress

Stress causes the distortion of semiconductor microstructures and results in changes in band structure. In the deformation potential theory, the strains are considered to be relatively small. The change in energy of each carrier subvalley, caused by the deformation of the lattice, is a linear function of the strain. By default (for silicon), Sentaurus Device (SDevice) considers three subvalleys for electrons (which are applied to three twofold subvalleys in the conduction band) and two subvalleys for holes (which are applied to heavy-hole and light-hole subvalleys in the valence band). The number of carrier subvalleys can be changed in the parameter file. We considered the Hamiltonian proposed by Bir and Pikus [21], which is finally combined with the Hamiltonian of Luttinger and Kohn [22], and that allows us to determine the effective masses and lifting of degeneration of the band. The Hamiltonian may be expressed as

$$
H_{BP} = \begin{bmatrix}
P+Q & -S & R & 0 & -\dfrac{S}{\sqrt{2}} & \sqrt{2}R \\[2ex]
-S & P-Q & 0 & R & -\sqrt{2}Q & \sqrt{\dfrac{3}{2}}S \\[2ex]
R & 0 & P-Q & S & \sqrt{\dfrac{3}{2}}S & \sqrt{2}Q \\[2ex]
0 & R & S & P+Q & -\sqrt{2}R & -\dfrac{S}{\sqrt{2}} \\[2ex]
-\dfrac{S}{\sqrt{2}} & -\sqrt{2}Q & \sqrt{\dfrac{3}{2}}S & -\sqrt{2}R & P & 0 \\[2ex]
\sqrt{2}R & \sqrt{\dfrac{3}{2}}S & \sqrt{2}Q & -\dfrac{S}{\sqrt{2}} & 0 & P
\end{bmatrix}
\tag{4.15}
$$

with

$$
P = -qa_v Tr(\bar{\varepsilon}), \quad Q = -q\frac{b}{2}\left(\varepsilon_{xx}^2 + \varepsilon_{yy}^2 + \varepsilon_{zz}^2\right),
$$

$$
R = -qb\frac{\sqrt{3}}{2}\left(\varepsilon_{xx}^2 + \varepsilon_{yy}^2\right) - iqd\varepsilon_{xy}^2, \quad S = -qd(\varepsilon_{zx} - i\varepsilon_{yz})
$$

where the front coefficients b and d are the potentials of deformation of the valence band and are specific to material considered. These are shown for Si in Table 4.1 [26].

In unstrained Si, the heavy-hole (HH) and the light-hole (LH) bands are degenerate at the Γ point as shown in Figure 4.5. The eigen states at the Γ point split into two groups due to spin-orbit coupling and are classified by $J = 3/2$ and $J = 1/2$ and degenerated in to HH ($J = 3/2$, $M_J = \pm3/2$)

TABLE 4.1

Deformation Potential of the Valence Band of Si

	Theoretical (eV)	Experimental (eV)
Δ_0	—	0.04
a_v	2.46	1.80
b	−2.35	−2.10 ± 0.10
d	−5.32	−4.85 ± 0.15

Source: Kasper, E., and D. J. Paul, *Silicon Quantum Integrated Circuit*, Springer-Verlag, Berlin, Heidelberg, 2005.

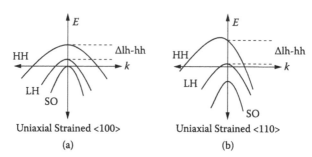

FIGURE 4.7
Simplified hole valance band structure for uniaxial strain in (a) <100> and (b) <110> directions.

and LH ($J = 3/2$, $M_J = \pm 1/2$) bands at the Γ point. The degeneracy between HH and LH bands is lifted due to application of strain on silicon crystal [27]. The Γ point at $k = 0$ has the full crystal symmetry. Uniaxial stress breaks the crystal symmetry by shortening (under compressive stress) and elongating (under tensile stress) and thus HH and LH bands are lifted. The HH and LH bands have negligible warping because they do not mix since the rotation symmetry is unchanged. Longitudinal uniaxial stress along <110> destroys the crystal symmetry more because the <100> axis has higher symmetry than the <110> axis. The lifting of energy bands in <100> and <110> directions is shown in Figure 4.7(a) and (b), respectively. VB-lifting: Due to the introduction of strain, the valence band warping for unstrained Si, Si under uniaxial compression is shown in Figure 4.7. The Si conduction band edges are located along the Δ valley and are sixfold degenerate.

Due to the strain induced by the uniaxial stress, the sixfold degenerate Δ valley splits into two valleys, i.e., Δ_2 and Δ_4 valleys, in either case. The splitting of energy and the ordering of the Δ_2 and Δ_4 valleys depend on the type and magnitude of the stress. Bir and Pikus [21] proposed another model for the strain-induced change in the energy of carrier subvalleys in silicon where they ignore the shear strain for electrons and suggest nonlinear dependence for holes:

$$\Delta E_{C,i} = \Xi_d \left(\varepsilon'_{11} + \varepsilon'_{22} + \varepsilon'_{33} \right) + \Xi_u \varepsilon'_{ii} \qquad (4.16)$$

$$\Delta E_{V,i} = -a \left(\varepsilon'_{11} + \varepsilon'_{22} + \varepsilon'_{33} \right) \pm \delta E \qquad (4.17)$$

where $\delta E = \sqrt{\frac{b^2}{2} \left((\varepsilon'_{11} - \varepsilon'_{22})^2 + (\varepsilon'_{22} - \varepsilon'_{33})^2 + (\varepsilon'_{11} - \varepsilon'_{33})^2 \right) + d^2 (\varepsilon'^2_{11} + \varepsilon'^2_{22} + \varepsilon'^2_{33})}$, and Ξ_d, Ξ_u, a, b, and d are other deformation potentials that correspond to the Bir and Pikus model. The sign ± separates heavy-hole and light-hole subvalleys of silicon in Equation (4.17). The above deformation potentials were used in our simulations. In the following, we present the parameters defined during simulation in the parameter file:

```
LatticeParameters
{
* DC2 (1) defines Bir & Pikus deformation potentials for
  conduction subband = 1
* DV2 (1) defines Bir & Pikus deformation potentials for
  valence subband = 1
* The subband energy shift due to strain (E) is equal to the
  following sum:
* (Bir & Pikus expression)
* D2 [1]*E11 + D2 [2]*E22 + D2 [3]*E33 +
* D2[4]*(0.5*D2[5]^2*((E11-E22)^2+(E22-E33)^2+(E33-E11)^2)+D2[
  6]*(E23^2+E13^2+E12^2))
*
* Egley's data for Bir & Pikus expressions:
*
* DC2 (1) = 9.5,0,0,0,0,0
* DC2 (2) = 0, 9.5,0,0,0,0
* DC2 (3) = 0, 0, 9.5,0,0,0
* DV2 (1) = 0, 0, 0,-1, 0.5,4
* DV2 (2) = 0, 0,0,1,0.5,4
*
  DC2 (1) = 9.5, 0.0000e+00, 0.0000e+00, 0.0000e+00,
   0.0000e+00, 0.0000e+00 # [eV]
  DC2 (2) = 0.0000e+00, 9.5, 0.0000e+00, 0.0000e+00,
   0.0000e+00, 0.0000e+00 # [eV]
  DC2 (3) = 0.0000e+00, 0.0000e+00, 9.5, 0.0000e+00,
   0.0000e+00, 0.0000e+00 # [eV]
  DV2 (1) = 0.0000e+00, 0.0000e+00, 0.0000e+00, -1.0000e+00,
   0.5, 4                  # [eV]
  DV2 (2) = 0.0000e+00, 0.0000e+00, 0.0000e+00, 1, 0.5, 4
                           # [eV]
}
```

Equations (4.16) and (4.17) have a common part $(\varepsilon_{11}' + \varepsilon_{22}' + \varepsilon_{33}')$, and therefore these expressions are combined in one general expression that gives a flexibility of its definition in the parameter file:

$$\Delta E_{B,i} = \delta E_1 + \delta E_2 + \delta E_3 \qquad (4.18)$$

where

$$\delta E_1 = \xi_{i1}^B (\varepsilon_{11}' + \varepsilon_{22}' + \varepsilon_{33}') + \xi_{i2}^B (\varepsilon_{11}' - \varepsilon_{33}') + \xi_{i3}^B (\varepsilon_{22}' - \varepsilon_{33}') + \xi_{i4}^B \varepsilon_{23}' + \xi_{i5}' \varepsilon_{13}' + \xi_{i6}' \varepsilon_{12}'$$

$$\delta E_2 = \xi_{i1}^{B2} \varepsilon_{11}' + \xi_{i2}^{B2} \varepsilon_{22}' + \xi_{i3}^{B2} \varepsilon_{33}'$$

$$\delta E_3 = \xi_{i4}^{B2} \sqrt{\frac{\left(\xi_{i5}^{B2}\right)^2}{2} \left((\varepsilon_{11}' - \varepsilon_{22}')^2 + (\varepsilon_{22}' - \varepsilon_{33}')^2 + (\varepsilon_{11}' - \varepsilon_{33}')^2 \right) + \left(\xi_{i6}^{B2}\right)^2 \left(\varepsilon_{11}'^2 + \varepsilon_{22}'^2 + \varepsilon_{33}'^2 \right)}$$

ξ_{ij}^{B2} are deformation potentials that correspond to the Bir and Pikus model, and ξ_{i4}^{B2} is a unitless constant that defines mainly a sign.

Using the stress tensor $\vec{\sigma}$ from the input file, Sentaurus Device computes from the stress coordinate system the tensor $\vec{\sigma}'$ in the crystal system using Equation (4.4). The strain tensor $\vec{\varepsilon}'$ is a result of applying Hooke's law in Equation (4.2) to the stress. Using Equations (4.16) and (4.17), or Equation (4.18), the energy band change can be computed for each conduction and valence carrier subvalleys. The modifications of the band structure when silicon is subjected to a stress have been discussed above. In the following section, we discuss the mobility models developed and used in device simulation.

4.8 Piezoresistive Mobility Model

In this mobility modelling, two types of piezoresistance effects are considered. One is the longitudinal piezoresistance coefficient $(\pi_{||})$, when the current and field are in the same direction of stress, and the other is transverse piezoresistance coefficient $(\pi_{\perp})$, when the current and field are perpendicular to stress. $(\pi_{||})$ and $(\pi_{\perp})$ for any arbitrary crystal orientation can be expressed as

$$\pi_{||} = \pi_{11} - 2(\pi_{11} - \pi_{12} - \pi_{44})\left(l_1^2 m_1^2 + m_1^2 n_1^2 + n_1^2 l_1^2\right) \tag{4.19}$$

$$\pi_{\perp} = \pi_{12} + 2(\pi_{11} - \pi_{12} - \pi_{44})\left(l_1^2 l_2^2 + m_1^2 m_2^2 + n_1^2 n_2^2\right) \tag{4.20}$$

where $l_i, m_i, n_i, (i = 1, 2, 3)$ and $(\pi_{\lambda,\mu})$ $(\lambda, \mu = 1, 2, 3)$ are the direction cosines and the components of piezoresistance tensor. External strain leads to a change in the effective masses and anisotropic scattering. The scattering of the electron and hole by deformation potential is considered with the aid of the phonon concept. Charge carriers colliding with phonons exchange energy and momentum with it. We obtained an expression for relaxation time $\tau(k)$ in terms of energy E as [28]

$$\tau(\varepsilon) = \frac{(\pi)^2 \hbar^4 S_{ij}}{2(2m^*)^{3/2} \Xi^2 N} \frac{E^{-1/2}}{k_B T} \tag{4.21}$$

where T is the absolute temperature. Relaxation time is proportional to $m^{*\pm 3/2}$ and directly proportional to the elasticity constant modulus S_{ij}. Elasticity modulus S_{ij} and deformation potential Ξ are specified in the parameter file. In τ approximation components σ_{ij} of conductivity tensor can be written as [29]

$$\sigma_{ij} = -\frac{e^2}{4\pi^3} \int \frac{\partial f_0(\varepsilon)}{\partial \varepsilon} \tau(\varepsilon) v_i v_j \delta \vec{k} \quad i, j \in \{x, y, z\} \tag{4.22}$$

where e is the electron charge, and $v_i = \hbar^{-1} \partial \varepsilon(\vec{k}, \hat{X})/\partial k_i$ is the ith component of the group velocity of charge carriers. The change in conductivity under stress is given by

$$\Delta \sigma_{ij} = -\frac{e^2}{4\Pi^3} \int \Delta \left[\frac{\partial f_0(\varepsilon)}{\partial \varepsilon} \tau(\varepsilon) v_i v_j \right] \delta\vec{k} \qquad (4.23)$$

Energy dependence of relaxation time $\tau(\varepsilon)$, i.e., Equation (4.21), is used to solve Equations (4.22) and (4.23). We use the first-order piezoresistance coefficients, which are determined by the relation

$$\pi_{ijkl} = -\frac{1}{\sigma_{ij}(\hat{X} = 0)} \left. \frac{\Delta \sigma_{ij}(\hat{X})}{\Delta X_{kl}} \right|_{\hat{X}=0}, \qquad i,j,k,l \in \{x,y,z\} \qquad (4.24)$$

External strain leads to a change in the effective masses and anisotropic scattering. The first effect is described by an independent constant term $\pi^{\alpha}_{ij,kon}$, but the second effect, the scattering, is calculated [30] at room temperature for low-doping concentrations $(\pi^{\alpha}_{ij,var})$ and multiplied by a doping-dependent and temperature-dependent factor $P_{\alpha}(N,T)$. Both effects are considered in the piezoresistive coefficients [30] as

$$\pi^{\alpha}_{ij} = \pi^{\alpha}_{ij,var} P_{\alpha}(N,T) + \pi^{\alpha}_{ij,kon} \qquad (4.25)$$

In case of electrons, scalar mobility used in the drift diffusion and hydrodynamic equations is a mean value averaged over the different conduction band minima. If the symmetry of crystal is destroyed, for example, by external strain, the conduction band valleys shift, and therefore electron transfer between the valleys occurs. This redistribution of electrons in the conduction band leads to anisotropic scattering. In the case of holes, the mobility is an averaged quantity including heavy and light holes. External strain leads to a lift of the degeneracy at the valence band maximum. The doping and temperature-dependent factor $P_{\alpha}(N,T)$ can be expressed as [30]

$$P_{\alpha}(N,T) = \frac{300}{T} \frac{F'_{1/2}\left(\dfrac{E_{F,\alpha}}{kT} \right)}{F_{1/2}\left(\dfrac{E_{F,\alpha}}{kT} \right)} \qquad (4.26)$$

where $F_{1/2}(x)$ and $F'_{1/2}(x)$ are the Fermi integrals of the order ½ and its first derivative. The Fermi energy E_F is equal to F_n-E_C for electrons and E_V-F_P for holes. They are calculated using appropriate analytic approximations [31] where the charge neutrality is assumed between carrier and doping (N), and it gives the doping dependence of the model. The numeric evaluation of $P_\alpha(N,T)$ is based on an analytic fit of the Fermi integrals [32]. Finally, the expression for piezoresistive coefficients is given by

$$\pi_{ij} = \pi_{ij,kon} + \pi_{ij,var} \cdot \frac{300}{T} \frac{F'_{s+(1/2)}(E_F/k_BT)}{F_{s+(1/2)}(E_F/k_BT)} \tag{4.27}$$

The piezoresistive model is applied in our simulations by including the name of the model in the subsection of the input command file. The effect of mechanical stress on the mobility may then be expressed in terms of the piezoresistive coefficient as follows:

$$\frac{\Delta\mu}{\mu} \approx \left|\pi_{\parallel}|\sigma_{\parallel} + \pi_\perp \sigma_\perp\right. \tag{4.28}$$

Using Equation (4.28), we have computed the mobility and subsequently simulated the MOSFET device characteristics. When dealing with the simulation of stress effect in silicon, there are three coordinate systems: the crystal system, the simulation system, and the stress system. Miller indices are used to describe the orientation of one system with respect to another. The simulation system with respect to the crystal system was defined in the parameter file **<file name>.par** in the **LatticeParameter** section, the default orientation of the simulation system. For the <110> channel direction in CMOS, the following parameter is used in the **<filename>.par** file [8].

```
LatticeParameters {
      X = (1, 0, 1)
      Y = (0, 1, 0) }
```

The orientation of the stress system with respect to the simulation system was defined in the **Device** command file, within the **Piezo** statement of the **Physics** section. The piezoresistive model was applied in simulation by including the name of the model in the subsection **Model** of the **Piezo** section of the input command file. With the specification of the piezoresistive coefficients, this section appears as follows:

```
Physics {
      Piezo (Model (...............))

      PiezoNkon = (π^n_{11,kon}, π^n_{12,kon}, π^n_{44,kon})

      PiezoNvar = (π^n_{11,var}, π^n_{12,var}, π^n_{44,var})

      PiezoPkon = (π^p_{11,kon}, π^p_{12,kon}, π^p_{44,kon})

      PiezoPvar = (π^p_{11,var}, π^p_{12,var}, π^p_{44,var})
      )
}
```

TABLE 4.2

Piezoresistive Parameters Used in Simulation for Electrons and Holes at 300 K

1×10^{-12} cm^2 dyn^{-1}	<100>		<110>	
Polarity	$\pi_\parallel$	$\pi_\perp$	$\pi_\parallel$	$\pi_\perp$
n or p	$\pi_{11,var}$	$\pi_{12,var}$	$(\pi_{11,var}+\pi_{12,var}+\pi_{44,var})/2$	$(\pi_{11,var}+\pi_{12,var}-\pi_{44,var})/2$
n-type	−102.6	53.4	−31.4	−17.8
p-type	1.5	1.5	56.5	−53.5
n or p	$\pi_{11,kon}$	$\pi_{12,kon}$	$(\pi_{11,kon}+\pi_{12,kon}+\pi_{44,kon})/2$	$(\pi_{11,kon}+\pi_{12,kon}-\pi_{44,kon})/2$
n-type	0	0	0	0
p-type	5.1	−2.6	15.25	−12.75

Piezoresistive parameters were incorporated in simulation for electrons and holes at 300 K and are given in Table 4.2.

In Sentaurus Device, there is a **DeformationPotential** statement in the **Piezo (Model (.................))** statement of the **Physics** section of the Device command file. The documentation states that **DeformationPotential** is used to reflect the stress effect on band structure, which coincides with the crystal system. Therefore, the default channel direction is <100>, not <110>.

4.9 Strain-Induced Mobility Model

Sentaurus Device has a built-in model for the mobility changes due to the carrier redistribution between subvalleys in Si [33]. As an example, the electron mobility is enhanced in a strained Si layer grown on top of a thick, relaxed SiGe. Due to the lattice mismatch (which can be controlled by the Ge mole fraction), the thin silicon layer appears to be stretched (under biaxial tension). The origin of the electron mobility enhancement can be explained by considering the sixfold degeneracy in the conduction band. The biaxial tensile strain lowers two perpendicular valleys (Δ_2) with respect to the fourfold in-plane valleys (Δ_4). Therefore, electrons are redistributed between valleys and Δ_2 is occupied more heavily. It is known that the perpendicular effective mass is much lower than the longitudinal one. Therefore, this carrier redistribution and reduced intervalley scattering enhance the electron mobility. The hole depends on the strain mainly due to redistribution of holes between light and heavy valleys, and changes the effective masses in these valleys. In the crystal coordinate system, the model gives only the diagonal elements of the electron mobility matrix, but for holes, the mobility is still isotropic. The following expressions have been proposed for the electron and hole mobilities [33]:

$$\mu_{ii}^n = \mu_n^0 \left[1 + \frac{1 - m_{nl}/m_{nt}}{1 + 2(m_{nl}/m_{nt})} \left(\frac{F_{1/2}\left(\frac{F_n - E_C - \Delta E_{C,i}}{kT} \right)}{F_{1/2}\left(\frac{F_n - E_C - \Delta E_C}{kT} \right)} - 1 \right) \right] \tag{4.29}$$

$$\mu^P = \mu_P^0 \left[1 + \left(\frac{\mu_{Pl}^0}{\mu_P^0} - 1 \right) \frac{(m_{Pl}/m_{Ph})^{1.5}}{1 + (m_{Pl}/m_{Ph})^{1.5}} \left(\frac{F_{1/2}\left(\frac{E_V + \Delta E_{V,l} - F_P}{kT} \right)}{F_{1/2}\left(\frac{E_V + \Delta E_{V,h} - F_P}{kT} \right)} - 1 \right) \right] \tag{4.30}$$

where μ_n^0 and μ_P^0 are electron and hole mobility models without the strain. m_{nl} and m_{nt} are the electron longitudinal and transfer masses in the subvalley. $\Delta E_{C,i}$ and ΔE_C are computed by Equations (4.29) and (4.30), respectively. The index i corresponds to a direction (for example, μ_{11}^n is the electron mobility in the direction of the X axis of the crystal system, and therefore $\Delta E_{C,1}$ should correspond to the twofold subvalley along the X axis). μ_{Pl}^0 is the mobility of light holes without the strain. m_{Pl} and $m_{P,h}$ are the hole light and heavy masses. $\Delta E_{V,l}$ and $\Delta E_{V,h}$ are computed also by Equation (4.18), with the specification of light-hole and heavy-hole subvalley numbers in the parameter file. F_n and F_p are quasi-Fermi levels of electrons and holes.

4.9.1 Strain-Induced Mobility Model under Electron–Phonon Interaction

In developing the strain-induced mobility model, we implemented strain effects by considering scattering of mobile charges in process-induced strained (PSS) n- and p-MOSFETs due to electron/hole-phonon interactions in the strained Si channel. To obtain the strained-induced mobility, first we calculated strain-induced interaction potential scattering by acoustic phonon. Then we used Fermi's golden rule to obtain the electron/hole-phonon scattering rates. Toward this, one needs to obtain the matrix element for electron–phonon scattering. The matrix element describes the coupling between initial and final electronic states due to interactions with scattering charge centres. Finally, we integrate the matrix elements over all final states to obtain the scattering rate between electron/hole-phonons.

4.9.2 Strain-Induced Interaction Potential Scattering by Acoustic Phonon

In a deformed Si substrate the coordinates of its lattice point are displaced. If the radius vector of a lattice point in undeformed condition is r, and in deformed condition is r', then the displacement vector is given by [34]

$$u(r) = r' - r \tag{4.31}$$

The deformation may be described in terms of a symmetrical strain tensor as

$$u_{im} = \frac{1}{2}\left(\frac{\partial u_i}{\partial u_m} + \frac{\partial u_m}{\partial u_i}\right) = u_{mi} = u^i \delta_{im} \tag{4.32}$$

If the distance between two lattice points of an undeformed Si substrate is dl and for a deformed Si substrate (strained Si substrate) is dl', then we get

$$dl^2 = dx^2 + dy^2 + dz^2 = \sum_{i=1}^{3} dx_i^2 \tag{4.33}$$

and

$$dl'^2 = \sum_{i=1}^{3} dx_i'^2 = \sum_{i=1}^{3}(dx_i + du_i)^2 = \sum_{i=1}^{3}(dx_i^2 + 2dx_i du_i + du_i^2) \tag{4.34}$$

The tensor du_i determines the variation of distances between the lattice points. Neglecting the quantity du_i^2 and expressing du_i in terms of du_m,

$$du_i = \sum_{m=1}^{3} \frac{\partial u_i}{\partial u_m} du_m \tag{4.35}$$

and combining Equations (4.33 to 4.35), we obtain dl'^2 as

$$dl'^2 = \sum_{i=1}^{3} dx'^2 = dl^2 + 2\sum_{i,m=1}^{3} \frac{\partial u_i}{\partial x_m} dx_i dx_m \tag{4.36}$$

Taking into account Equation (4.32) we obtain for dl'^2

$$dl'^2 = dl^2 + \sum_{i,m=1}^{3}\left(\frac{\partial u_i}{\partial x_m} + \frac{\partial u_m}{\partial x_i}\right)dx_i dx_m = dl^2 + 2\sum_{i,m} u_{im} dx_i dx_m \tag{4.37}$$

We may write from Equation (4.37)

$$dl' = \sqrt{dl^2 + 2\sum_{i,m} u^i \delta_{im} dx_i dx_m} = \sqrt{dl^2 + 2\sum_{i} u^i dx_i^2} \tag{4.38}$$

The variation of volume upon deformation is given by

$$\Delta V' = dx' dy' dz' = dx\left(1 + u^{(1)}\right)dy\left(1 + u^{(2)}\right)dz\left(1 + u^{(3)}\right)$$

$$= \Delta V\left(1 + u^{(1)} + u^{(2)} + u^{(3)}\right) \tag{4.39}$$

Therefore, the local dilation Δr is given by

$$\Delta(r) = \frac{\Delta V' - \Delta V}{\Delta V} = \sum_i u^{(i)} = \sum_i u_{ii} = \sum_i \frac{\partial u_i}{\partial x_i} = \nabla.u(r) \qquad (4.40)$$

The above-mentioned local dilation relation is equivalent to displacement of the atoms, and hence equivalent to a local change of lattice parameter. Therefore, it induces a modification of both bands: (1) conduction band (E_C) and (2) valence band (E_V). The interaction potential H_{e-ph}^{AC} of the acoustic phonon with the lattice depends on the variation of the conduction band and valence band edges. Since phonons deform the crystal in three dimensions, for small stress and for an isotropic crystal, interaction potential is given by

$$H_{e-ph}^{AC} = \Xi \frac{\delta V}{V} = \Xi \nabla.u(r) \qquad (4.41)$$

where Ξ is the so-called deformation potential. The lattice displacement $u(r)$ for long wavelength phonon is given by [34]

$$\vec{u}(r) = \sum_q \vec{w}_q \left(\frac{\hbar}{2\rho\omega_q} \right)^{1/2} \left(a_q e^{(i\vec{q}.\vec{r})} + a_q^\dagger e^{(-i\vec{q}.\vec{r})} \right) \qquad (4.42)$$

where ρ is the semiconductor density and $\vec{w}_q$ is the polarisation vector. For longitudinal phonons, the polarisation vector is $\vec{w}_q = \frac{\vec{q}}{q} = \hat{q}$. Thus the acoustic deformation potential is written as

$$H_{e-ph}^{AC} = \Xi \frac{\delta V}{V} = \Xi \nabla.\vec{u}(r) = i \sum_q (\vec{w}_q.\vec{q}) \left(\frac{\hbar\Xi^2}{2\rho\omega_q} \right)^{1/2} \left(a_q e^{(i\vec{q}.\vec{r})} - a_q^\dagger e^{(-i\vec{q}.\vec{r})} \right) \qquad (4.43)$$

4.9.3 Transition Probability for Acoustic Phonon Scattering

The scattering rate of electrons or holes by lattice vibration in the presence of strain may be explained in terms of the corpuscular model with the aid of the phonon concept. Charge carriers colliding with phonons exchange energy and quasi-momentum with it. Since the number of phonons depends on temperature, the charge scattering should be temperature dependent. However, in order to calculate the quantum transitions of electrons and holes from state to state, the perturbation generated by deformation potential due to strain should be applied. Once we have the acoustic deformation potential, one may obtain the scattering rate using Fermi's golden rule:

$$\Gamma_{i\rightarrow f} = \frac{2\pi}{\hbar} \left| \left\langle f \left| H_{e-ph}^{AC} \right| i \right\rangle \right|^2 \delta(E_f - E_i) \qquad (4.44)$$

where f and i refer to final and initial states. The energy of lattice vibrations under electrons or holes phonon interaction may change by the creation or annihilation of a phonon. Hence, in collisions the initial and final energies of the electron/hole-phonon system are

$$
\left.\begin{array}{l}
E_i = E_k + n_q \hbar \omega_q \\
E_f = E_{k'} + n'_q \hbar \omega_q
\end{array}\right\} n'_q = n_q \pm 1
\tag{4.45}
$$

Hence, transition probability for the state E_k to $E_{k'}$ in an electron/hole-phonon collision involving a phonon in the wave vector q is

$$
\Gamma_{k \to k'} = \frac{2\pi}{\hbar} \left| \langle k' | H_{e-ph}^{AC} | k \rangle \right|^2 \delta(E_k - E_{k'} \pm \hbar \omega_q)
\tag{4.46}
$$

4.9.4 Strain-Induced Scattering Matrix

If $\psi_k(\vec{r})$ is the eigenfunction for the unstrained condition and $\psi_k(\vec{r}, \varepsilon)$ is the eigenfunction for the strained condition, then

$$
\psi(\vec{r}, \varepsilon) = u_{\vec{k}}(\vec{r}, \varepsilon) e^{i\vec{k}.\vec{r}} = \psi(\vec{r}) + \delta\psi(\vec{r})
\tag{4.47}
$$

Now the scattering matrix elements for long-wavelength acoustic phonon scattering in isotropic material are given as [28]

$$
\Im(k', k) = \langle k' | H_{e-ph}^{AC} | k \rangle
$$
$$
= i \sum_q (\vec{w}_q . \vec{q}) \left(\frac{\hbar \Xi^2}{2\rho \omega_q} \right)^{1/2} \left[\int \psi_{k'}^* \phi_{n'_q}^* \left(a_q e^{(i\vec{q}.\vec{r})} - a_q^\dagger e^{(-i\vec{q}.\vec{r})} \right) \psi_k \phi_{n_q} dV \right]
\tag{4.48}
$$

Now, we consider the following:

$$
\psi_k = u_k(r) e^{i\vec{k}.\vec{r}}, \quad \psi_k = u_{k'}^*(r) e^{-i\vec{k}'.\vec{r}}
\tag{4.49}
$$

$$
a_q \phi_{n_q} = \sqrt{n_q} \phi_{n_q-1}, \quad a_q^\dagger \phi_{n_q} = \sqrt{(n_q + 1)} \phi_{n_q+1}
\tag{4.50}
$$

where $a_q^\dagger$ and a_q are the creation and annihilation operators, and ψ_k and ϕ_q are the wave functions of the electron and lattice vibration mode of vector $\vec{q}$. Now substituting Equations (4.49) and (4.50) in Equation (4.48), one obtains

$$
\Im(k', k) = i \sum_q (\vec{w}_q . \vec{q}) \left(\frac{\hbar \Xi^2}{2\rho \omega_q} \right)^{1/2} \int \left[\begin{array}{l} u_{k'}^* \phi_{n'_q}^* u_k(r) \phi_{n_q-1} \sqrt{n_q} e^{\{i(\vec{k}+\vec{q}-\vec{k}').\vec{r}\}} \\ - u_{k'}^* \phi_{n'_q}^* u_k(r) \phi_{n_q+1} \sqrt{n_q + 1} e^{\{i(\vec{k}-\vec{q}-\vec{k}').\vec{r}\}} \end{array} \right] dV
\tag{4.51}
$$

4.9.5 Relaxation Time for Acoustic Phonon Scattering

The well-known relation to describe relaxation time is given by [35]

$$\frac{1}{\tau} = \frac{V}{(2\pi)^3} \int \left(1 - \frac{k'\cos\theta'}{k\cos\theta}\right) \Gamma_{k\to k'} dk' \tag{4.52}$$

Replacing and combining Equations (4.46), (4.48), and (4.52), we get the following:

$$\frac{1}{\tau} = \frac{1}{(2\pi)^3}\left(\frac{\pi\Xi^2}{\rho}\right) \int \left[\frac{(n_q+1)}{\omega_q}q^2\delta\left(E_{k'}-E_k+\hbar\omega_q\right) + \frac{n_q}{\omega_q}q^2\delta\left(E_{k'}-E_k-\hbar\omega_q\right)\right]$$

$$\times \left(1 - \frac{k'\cos\theta'}{k\cos\theta}\right) dk' \tag{4.53}$$

where n_q represents the occupation number of acoustic phonons with wave vector $\vec{q}$. Since an electron or hole may change its state from **k** by emission or absorption of an acoustic phonon of energy $\hbar\omega_q$, there are two terms involved in $\Gamma_{k\to k'}$ corresponding to these two types of transitions. Since $\vec{k}' = \vec{k} + \vec{q}$, the integral may also be carried out in **q** space. τ may therefore be written as

$$\frac{1}{\tau} = \frac{1}{(2\pi)^3}\left(\frac{\pi\Xi^2}{\rho}\right) \int \left[(n_q+1)\delta(E_{k'}-E_k+\hbar\omega_q) + n_q\delta(E_{k'}-E_k-\hbar\omega_q)\right]$$

$$\times \frac{q^2}{\omega_q}\left(1 - \frac{k'\cos\theta'}{k\cos\theta}\right) d\vec{q} \tag{4.54}$$

The element of volume $d\vec{q}$ in **q** space may be expressed in the spherical coordinate system as shown in Figure 4.8.

$$d\vec{q} = q^2 \sin\beta d\beta d\phi dq \tag{4.55}$$

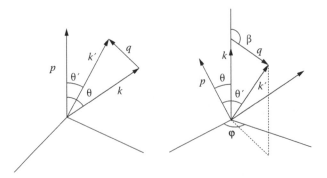

FIGURE 4.8
Reference coordinate system.

Thus, we get

$$
\frac{1}{\tau} = \frac{1}{(2\pi)^3} \left(\frac{\pi \Xi^2}{\rho} \right) \int_{q_{min}}^{q_{max}} \int_{-1}^{1} \int_{0}^{2\pi} \left[\begin{array}{c} (n_q + 1)\delta\left(E_{k'} - E_k + \hbar\omega_q\right) \\ + n_q\delta\left(E_{k'} - E_k - \hbar\omega_q\right) \end{array} \right]
$$

$$
\times \frac{q^2}{\omega_q} \left(1 - \frac{k'\cos\theta'}{k\cos\theta} \right) q^2 dq d(\cos\beta) d\phi \tag{4.56}
$$

For the uniaxial strain-induced transformation of the isoenergetic surface, the deformed spheres transform into the oblate ellipsoid in the heavy-hole band and elongated ellipsoid in the light-hole band [36]. For this case Ξ is a function of β, i.e., $\Xi(\beta)$, and

$$
\delta\left(E_{k'} - E_k \mp \hbar\omega_q\right) = \delta\left(\frac{\hbar^2 k'^2}{2m^*} - \frac{\hbar^2 k^2}{2m^*} \mp \hbar vq \right) = \delta\left(\frac{\hbar^2 k'^2}{2m^*} \pm \hbar vq \cos\theta \mp \hbar vq \right) \tag{4.57}
$$

For the azimuthal average approximation we can take $\omega_q = vq$, where v is velocity of the mode averaged over direction. Regarding n_q, a simple case is considered and the most common application is the case of equipartion, which is given by

$$
n_q = \frac{1}{e^{\hbar\omega_q/k_B T} - 1} \approx \frac{k_B T}{\hbar\omega_q} \qquad \frac{k_B T}{\hbar\omega_q} \ll 1 \tag{4.58}
$$

Since there is an energy and momentum conservation limit, we take $q_{min} = 0$ to $q_{max} = 2k$; a typical phonon energy is $\hbar vk$. When $n_q \gg 1$, the rates for absorption and emission become almost identical. Substituting Equations (4.56) and (4.57) in Equation (4.55) and using the above approximations for a spherical band, one obtains

$$
\frac{1}{\tau} = \frac{k_B T}{8\pi^2 \hbar\rho v^2} \int_{q_{min}}^{q_{max}} \int_{-1}^{1} \int_{0}^{2\pi} \Xi^2(\beta)(E_{k'} - E_k \mp \hbar\omega_q)q^2 dq d(\cos\beta)d\phi \tag{4.59}
$$

However, for semiconductors having ellipsoidal constant energy surfaces (for uniaxial strain), the matrix element for the acoustic phonon scattering is not isotropic. It is found that the relaxation time for this case may be expressed in two components, one perpendicular to the axis of symmetry of band structure and the other parallel to it. Using the relation

$\rho v^2 = S_{ij}$, where S_{ij} is the elasticity constant modulus, the components may be expressed as

$$
\left.
\begin{aligned}
\frac{1}{\tau_{\|}} &= \frac{3\pi}{S_{ii}} \frac{m_D^{3/2}}{2^{3/2}} \frac{kTE_k^{1/2}}{\pi^2\hbar^4}\left(\xi_{\|}\Xi_d^2 + \eta_{\|}\Xi_d\Xi_u + \zeta_{\|}\Xi_u^2\right) \\[2ex]
\frac{1}{\tau_{\perp}} &= \frac{3\pi}{S_{ii}} \frac{m_D^{3/2}}{2^{3/2}} \frac{kTE_k^{1/2}}{\pi^2\hbar^4}\left(\xi_{\perp}\Xi_d^2 + \eta_{\perp}\Xi_d\Xi_u + \zeta_{\perp}\Xi_u^2\right)
\end{aligned}
\right\}
\tag{4.60}
$$

where S_{ii} now represents an average elastic constant for longitudinal waves; $\xi_{\|}, \eta_{\|}, \zeta_{\|}, \xi_{\perp}, \eta_{\perp}$, and $\zeta_{\perp}$ are dimensionless constants; Ξ_d is the deformation potential constant for dilation; and Ξ_u is that for a uniaxial strain. m_D is the density of states effective mass for the ellipsoidal energy surface.

4.10 Implementation of Mobility Model

Elasticity modulus S_{ij} (10^{12} dyn/cm^2) is specified in the field $S_{[i][j]}$ in the parameter file. The values of S_{11}, S_{12}, and S_{44} are 1.23×10^{12}, -4.76×10^{12}, and 0.8×10^{12}, respectively [8]. The total deformation potential constants (Ξ) for conduction and valance bands were taken as 9.5 and 6.6 eV, respectively [20, 27]. For the case with 500 MPa uniaxially compressive stresses in Si, E_k was assumed to be 25 meV. Scattering by neutral centre and scattering by impurity ion were also considered in simulation. As all the mechanisms are independent of each other, the total scattering probability is equal to the sum of probabilities of scattering by scattering centres of all types. Hence, the mobility model is given by

$$
\mu_r = \langle\tau\rangle = \frac{e\langle\tau\rangle}{m_D^*} = \frac{e}{m_D^*}\left\langle \frac{1}{\displaystyle\sum \frac{1}{\tau_i(E)}} \right\rangle
\tag{4.61}
$$

The above mobility was considered in a hydrodynamic model and was implemented in the Sentaurus Device simulator. To activate the mobility model, appropriate mobility values were defined in the fields of the parameter file of the device simulator. Simulated hole mobility for process-induced strained Si p-MOSFET is shown in Figure 4.9. As expected, higher hole mobility is seen in the direction <110>.

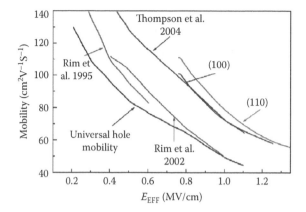

FIGURE 4.9
Strained Si hole mobility enhancement vs. vertical electric field (SiGe S/D, biaxial substrate stress, and HOT).

4.11 Summary

Stress–strain engineering is now a key performance booster for CMOS devices. Mechanical stress can affect band gap, effective mass, and carrier mobility. The band structure of strained Si considering spin-orbit coupling and the k.p method has been considered to get a deeper understanding about the energy band. The piezoresistance mobility model, stress-induced mobility model, and deformation potential-based electron/hole-phonon interaction-based mobility model have been developed. Procedures for implementing the models in Sentaurus Device have been described in detail.

Review Questions

1. What is mobility scaling?
2. What is a piezoresistive model?
3. What is a valence band offset in SiGe?
4. What is a conduction band offset in SiGe?
5. What is type II band alignment?
6. What are heavy- and light-hole bands?
7. How can band structures be modified by applied stress?
8. What is the role of acoustic phonons on mobility?

9. What is a scattering matrix?

10. What is the role of electron–phonon interaction on mobility?

11. How does splitting of energy subbands affect electron mobility?

References

1. J. Bardeen and W. Shockley, Deformation Potentials and Mobilities in Non-Polar Crystals, *Phys. Rev.*, 80, 72–80, 1950.
2. C. S. Smith, Piezoresistance Effect in Germanium and Silicon, *Phys. Rev.*, 94, 42–49, 1954.
3. V. M. Agostinelli, H. Shin, and A. F. Tasch, A Comprehensive Model for Inversion Layer Hole Mobility for Simulation of Submicrometer MOSFETs, *IEEE Trans. Electron Dev.*, 38, 151–159, 1991.
4. M. Shirahata, H. Kusano, N. Kotani, S. Kusanoki, and Y. Akasaka, A Mobility Model Including the Screening Effect in MOS Inversion Layer, *IEEE Trans. Computer-Aided Design*, 11, 1114–1119, 1992.
5. M. N. Darwish, J. L. Lentz, M. R. Pinto, P. M. Zeitzoff, T. J. Krutsick, and H. H. Vuong, An Improved Electron and Hole Mobility Model for General Purpose Device Simulation, *IEEE Trans. Electron Dev.*, 44, 1529–1538, 1997.
6. M. Kondo and H. Tanimoto, An Accurate Coulomb Mobility Model for MOS Inversion Layer and Its Application to NO Oxynitride Devices, *IEEE Trans. Electron Dev.*, 48, 265–270, 2001.
7. S. Reggiani, A. Valdinoci, L. Colalongo, M. Rudan, G. Baccarani, A. Stricker, F. Illien, N. Felber, W. Fichtner, S. Mettler, S. Lindenkreuz, and L. Zullino, Surface Mobility in Silicon at Large Operating Temperature, Proc. Simulation Semiconductor Processes Devices (SISPAD), 15–20, 2002.
8. Synopsys, Inc., *Sentaurus Device User Manual*, version A-2007.12, Mountain View, CA, March 2008.
9. T. K. Maiti, S. S. Mahato, P. Chakraborty, S. K. Sarkar, and C. K. Maiti, CMOS Performance Enhancement in Hybrid Orientation Technologies, *J. Comput. Electron*, 7, 181–186, 2008.
10. J. F. Nye, *Physical Properties of Crystals*, Clarendon Press, Oxford, 1985.
11. Synopsys, Inc., *Sentaurus Process User Manual*, version A-2007.12, Mountain View, CA, March 2008.
12. H. E. Randell, Application of Stress from Boron Doping and Other Challenges in Silicon Technology, PhD thesis, University of Florida, 2005.
13. M. S. Lundstrom, *Fundamentals of Carrier Transport*, Cambridge University Press, Cambridge, 2000.
14. M. S. Lundstrom and Z. Ren, Essential Physics of Carrier Transport in Nanoscale MOSFETs, *IEEE Trans. Electron Dev.*, 49, 131–141, 2002.
15. ST Microelectronics, *MASTAR User Guide*, 2003.
16. M. V. Fischetti and Z. Ren, Six-Band k.p Calculation of the Hole Mobility in Silicon Inversion Layers: Dependence on Surface Orientation, Strain, and Silicon Thickness, *J. Appl. Phys.*, 94, 1079–1095, 2003.

17. J. R. Chelikowsky and M. L. Cohen, Nonlocal Pseudopotential Calculations for the Electronic Structure of Eleven Diamond and Zincblende Semiconductors, *Phys. Rev. B*, 14, 556–582, 976.

18. C. G. Van de Walle and R. M. Martin, Theoretical Calculations of Heterojunction Discontinuities in the Si/Ge System, *Phys. Rev. B*, 34, 5621–5634, 1986.

19. M. M. Rieger and P. Vogl, Electronic-Band Parameters in Strained Si Ge Alloys on Si Ge Substrates, *Phys. Rev. B*, 48, 14276–14287, 1993.

20. M. V. Fischetti and S. E. Laux, Band Structure, Deformation Potentials, and Carrier Mobility in Strained Si, Ge, and SiGe Alloys, *J. Appl. Phys.*, 80, 2234–2252, 1996.

21. G. L. Bir and G. E. Pikus, *Symmetry and Strain Induced Effects in Semiconductors*, Wiley, New York, 1974.

22. J. M. Luttinger and W. Kohn, Motion of Electrons and Holes in Perturbed Periodic Fields, *Phys. Rev.*, 97, 869–883, 1955.

23. J. M. Luttinger, Quantum Theory of Cyclotron Resonance in Semiconductors: General Theory, *Phys. Rev.*, 102, 1030–1041, 1956.

24. S. E. Thompson, M. Armstrong, C. Auth, S. Cea, R. Chau, G. Glass, T. Hoffman, J. Klaus, Z. Ma, M. Bohr, and Y. El-Mansy, A 90-nm Logic Technology Featuring Strained-Silicon, *IEEE Trans. Electron Dev.*, 51, 1790–1797, 2004.

25. T. Skotnicki, J. A. Hutchby, T.-J. King, H. S. P. Wong, and F. Boeuf, The End of CMOS Scaling: Toward the Introduction of New Materials and Structural Changes to Improve MOSFET Performance, *IEEE Circuits Devices Mag.*, 21, 16–26, 2005.

26. E. Kasper and D. J. Paul, *Silicon Quantum Integrated Circuit*, Springer-Verlag, Berlin, 2005.

27. Y. Sun, S. E. Thompson, and T. Nishida, Physics of Strain Effects in Semiconductors and Metal-Oxide-Semiconductor Field-Effect Transistors, *J. Appl. Phys.*, 101, 104503-1–104503-22, 2007.

28. P. Roblin and H. Rohdin, *High-Speed Heterostructure Devices*, Cambridge University Press, Cambridge, 2002.

29. K. Matsuda, K. Suzuki, K. Yamamura, and Y. Kanda, Nonlinear Piezoresistance Effects in Silicon, *J. Appl. Phys.*, 73, 1838–1847, 1993.

30. Y. Kanda, A Graphical Representation of the Piezoresistance Coefficients in Silicon, *IEEE Trans. Electron Dev.*, ED-29, 64–70, 1982.

31. S. Selberherr, *Analysis and Simulation of Semiconductor Devices*, Springer, Wien, 1984.

32. S. Wolfram, *Mathematica: A System for Doing Mathematics by Computer*, 2nd ed., Addison-Wesley Publishing Company, Redwood City, CA, 1991.

33. J. L. Egley and D. Chidambarrao, Strain Effects on Device Characteristics: Implementation in Drift-Diffusion Simulators, *Solid-State Electron.*, 36, 1653–1664, 1993.

34. C. Kittel, *Quantum Theory of Solid*, John Wiley & Sons, New York, USA, 1987.

35. B. R. Nag, *Physics of Quantum Well Devices*, Kluwer Academic Publishers, Boston, USA, 2000.

36. V. Kolomoets, V. Baidakov, A. Fedosov, A. Gorin, V. Ermakov, E. Liarokapis, G. Gromova, B. Kazbekova, L. Taimuratova, and B. Orasgulyev, Application of Piezoresistance Effect in Highly Uniaxially Strained p-Si and n-Si for Current-Carrier Mobility Increase, *Phys. Stat. Sol. B*, 246, 652–654. 2009.

5

Strain-Engineered MOSFETs

From the complementary metal-oxide-semiconductor (CMOS) technology node beyond 90 nm, it has become very difficult to improve device performance by only reducing the physical gate length. According to the International Technology Roadmap for Semiconductors [1], by the year 2015, the channel length of metal-oxide-semiconductor field-effect transistors (MOSFETs) is projected to be less than 10 nm. The historic performance enhancement trend can probably continue until the 11 nm node with physical gate length no shorter than 10 nm. The logic technology node and physical gate length as a function of year of introduction are shown in Figure 5.1. For instance, at the technology nodes of 130 and 90 nm, the physical gate lengths are reduced to ~70 and ~50 nm, respectively. At the end of this decade, the difference in the physical gate length and the technology node could reach as much as 50%. It may be noted that beyond the 130 nm node, the scale of the physical gate length has entered into the nanometer regime. The challenge of fabricating such gate length lies in a much higher level of integration. There have been reports suggesting that the fundamental limit of scaling is at or near a gate length of 25 nm. Following Moore's law is becoming extremely difficult for the upcoming technology nodes, where the main challenging point for device scaling is the off-state leakage current. Planar MOSFETs with gate lengths as short as 5 nm have been fabricated; however, owing to huge off-state currents, they are not suitable for future integrated circuits (ICs).

According to the scaling theory, for every reduction in the transistor size, a corresponding decrease in the power supply voltage is required. From the device integration point of view, however, this extreme scaling of transistors degrades performance, which is contradictory to the objective of the scaling theory. For gate lengths larger than 100 nm, drain current improvement is expected following Moore's law. In contrast, when the gate length is reduced into the sub-100 nm range, drain current improvement is lost. It is postulated that the direct tunneling leakage between the gate and the source and drain is the cause of this degradation [2]. Hence, introduction of deep sub-100 nm bulk Si transistors into the market remains uncertain if the integration issues are unresolved. This drawback has prompted research to explore other means to achieve performance enhancement in new CMOS technology generations. As the MOSFET channel length enters the nanometer regime, however, short-channel effects (SCEs), such as threshold voltage roll-off and drain-induced barrier lowering (DIBL), become high, which hinders the

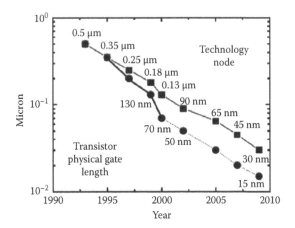

FIGURE 5.1
Logic technology node and physical gate length as a function of year of introduction. (After Maiti, T. K., Process-Induced Stress Engineering in Silicon CMOS Technology, PhD thesis, Jadavpur University, 2009.)

scaling capability of planar bulk or silicon-on-insulator (SOI) MOSFETs. The three primary SCEs are:

1. V_{th} roll-off: A portion of the channel is already depleted, and hence the gate electrode does not have to alter the potential at the dielectric interface near the source/drain junction as much to invert the channel V_{th}.

2. S/D off-state leakage: As the depletion width increases further into the body/channel area, a large V_{ds} results in carriers traversing through the body rather than the channel from source to drain, a phenomenon commonly referred to as punch-through I_{off} increases.

3. DIBL: As the drain bias is increased to ensure velocity saturation of carriers in the channel, the depletion region near the drain electrode creeps further into the channel and undermines or replaces gate control of the transistor, with drain control I_d becoming independent of V_g.

As channel length is scaled down, the performance is expected to increase as a result of the decrease in the intrinsic channel resistance. However, the extrinsic series resistance does not scale proportionately and is becoming a significant part of the total device resistance.

To overcome these problems, new device architectures as well as new gate stacks are being investigated. Multigate (also known as FinFET) devices are considered a promising architecture for replacement of conventional planar MOSFET devices, offering a solution for overcoming the short-channel effects and providing better V_{th} control at short gate lengths. The nonplanar architecture in combination with the small dimensions of the devices helps to improve the on-state drive current and reduce the off-state leakage current. These advantages translate to lower power consumption and enhanced device performance. Nonplanar devices are also more compact than conventional planar transistors, enabling higher transistor density, which translates to smaller overall microelectronics.

The limitation of the downscaling of single-gate, planar bulk MOSFETs is the inherent poor electrostatic control of the gate over the channel and poor transport properties of the carrier. Therefore, it is imperative to find a solution that encompasses materials to a new architecture for future technology nodes. Device structures are being scaled from 3D (bulk CMOS), quasi 2D (partially depleted SOI), 2D (fully depleted SOI), and quasi 1D (nanowire FET and tri-gate FinFET) for better channel electrostatics. To reduce the short-channel effects, researchers have proposed double-gate MOSFETs and FinFET devices. In tri-gate devices the gate is placed on the three sides of the channel. This results in a better control on the channel and significant reduction in the drain-to-source subthreshold leakage current. Many novel device structures and materials such as silicon nanowire transistors, carbon nanotube FETs, and molecular transistors have been proposed.

There are several ways by which the device engineer can handle the SCE, including reduction of S/D junction depth, increasing dopant concentration in the channel, and decreasing the effective oxide thickness (EOT) of the gate dielectric. Reducing EOT results in increased capacitive coupling between the gate electrode and channel region. This implies a greater ease in altering the surface potential at the dielectric/channel interface that leads to inversion. EOT scaling progressed for many years simply by reducing the thickness of the gate dielectric, thereby increasing its capacitance. In order to control short-channel effects in aggressively scaled MOSFETs, one must ensure also that the ratio of body thickness to gate length is sufficient to ensure both low off-state leakage and full gate control over the channel. Decreasing the S/D junction depth and increasing dopant concentration in the channel are the techniques that are almost exclusively applied in planar MOSFET fabrication. In order to suppress the SCE in bulk MOSFETs, other parameters need to be scaled down together with L_g, such as the gate oxide thickness (T_{ox}), the channel depletion width (X_d), and the source/drain junction depth (X_j). However, the thickness of SiO_2-based gate dielectrics is approaching physical limits (<2 nm), for which quantum mechanical tunneling induces severe gate leakage current through the gate dielectric. The off-state leakage current (I_{off}) increases as gate length (L_g) decreases because capacitive control of the channel potential by the gate becomes more difficult. Metal gate technology offers tunable work function for V_{th} adjustment and allows further MOSFET scaling because it eliminates the issues of poly-Si gate technology, namely, the gate depletion effect and boron penetration.

Recently there has been a great deal of interest in channel engineering through the introduction of local stress. Both the SiGe and SiC have been used in the source/drain regions to introduce stress locally for device drive current enhancement. Intel first introduced the new hafnium-based dielectrics with metal gates for its 45 nm technology node. Together with new dielectric and metal gates, an improved technique to induce more strain into the channel for obtaining enhanced performance was employed for the 22 nm technology node. Toward performance enhancement, the CMOS technology will utilise various approaches, such as advanced MOSFET structures, metal gate with tunable work function, strained Si, and channel orientation optimisation. In addition to these process-based solutions to control SCE for scaled device dimensions, there exist fundamental limitations on device size that will be addressed in the following sections. The objective of this chapter is performance evaluation and prediction for nanoscale devices in silicon technology beyond the 45 nm CMOS technology node.

5.1 Process Integration

As discussed above, some of the main challenges faced by the Si CMOS technology are large short-channel effects resulting in an exponential increase in leakage power, process variations resulting in large deviations in the performance of the circuits, and technological limitations. Leakage power is broadly classified into two categories: standby leakage, which corresponds to the situation when the circuit is in a nonoperating or sleep mode, and active leakage, which relates to leakage during normal operation. Technology boosters such as strain have helped the continuation of CMOS historic performance trend up to the 45 nm node. Process integration challenges are (1) power consumption, (2) leakage current, (3) metal gate electrodes, and (4) high-k gate dielectrics, which are discussed below.

5.1.1 Power Consumption

As transistor sizes shrink, the total power consumption of chips is becoming a dominant factor in determining the chip performance. The power consumption of microprocessor cores and interconnects constitutes a significant portion of the total power consumption of modern microprocessors. Leakage current is a primary concern for low-power, high-performance digital CMOS circuits for portable applications, and industry trends show that leakage will be the dominant component of power in future technologies. According to the International Technological Roadmap for Semiconductors, physical oxide thickness (T_{ox}) values of 7–12 Å will be required for high-performance CMOS circuits, and quantum effects that cause tunneling will play a dominant role in such ultra-thin oxide devices. Dynamic and leakage power has consistently increased with every technology generation. As VLSI technology scales, the enhanced performance of smaller transistors comes at the expense of increased power consumption. In addition to the dynamic power consumed by the circuits, there is a tremendous increase in the leakage power consumption, which increases with the increase in the operating temperature. One of the major challenges is the reduction of the supply voltage due to smaller gate oxides that cannot withstand the traditional 3.3 and 5 V supplies. Reducing the supply voltage necessitates the reduction of the transistor threshold to maintain an adequate overdrive voltage. In turn, the reduction of the threshold voltage increases the transistor's subthreshold conduction, which translates into an increase in leakage power consumption.

5.1.2 Leakage Current

In nanoscale CMOS devices, leakage power is the major contributor to total power consumption. When electrons tunnel into the conduction band of the oxide layer, it is called Fowler-Nordheim tunneling. When the oxide layer is very thin, say 4 nm or less, then instead of tunneling into the conduction band of the SiO_2 layer, electrons from the inverted silicon surface can tunnel directly through the forbidden energy gap of the SiO_2 layer. This is called direct tunneling. The leakage currents are illustrated in Figure 5.2. Various mechanisms that contribute to total leakage power in the short-channel devices are (1) the leakage current due to the reverse-bias p-n

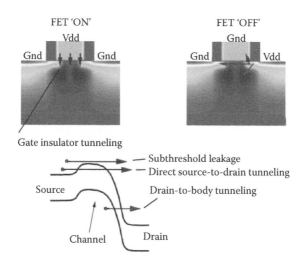

FIGURE 5.2
Leakage currents in an n-type MOSFET. When the FET is ON, the major leakage source is the gate tunneling current. When the FET is OFF, the major three leakage currents are shown in the band diagram. (After Liu, M., 10-nm CMOS—A Design Study on Technology Requirement with Power/Performance Assessment, PhD thesis, University of California, San Diego, 2007.)

junction, (2) leakage current due to the subthreshold leakage, (3) current due to the tunneling of carriers through the thin gate oxide, (4) current flowing in the gate because of an injection of hot carriers, (5) current because of gate-induced drain lowering (GIDL), and (6) current because of a channel punch-through. When the electric field across a reverse-biased p-n junction approaches 10^6 V/cm, significant current flow can occur due to tunneling of electrons from the valence band of the p-region into the conduction band of the n-region. The band-to-band tunneling poses as a limit to CMOS scaling. As CMOS channel length is scaled to 10 nm or below, the source-to-drain direct tunneling becomes the dominant leakage.

5.1.3 Metal Gate Electrodes

A significant advantage of employing a midgap metal arises from a symmetrical V_{th} value for both n- and p-MOSFETs, because by definition the same energy difference exists between the metal Fermi level and the conduction and valence bands of Si. In order to continue device scaling, high-k gate dielectrics are required, and metal gates show superior compatibility over silicon gates. The work function is an essential parameter in optimising electrical characteristics, specifically the threshold voltage. A polysilicon gate has the advantage that it can be doped p-type or n-type, shifting the work function so that it is suitable for n- and p-MOSFET devices, thereby simplifying integration. As gate oxide thickness decreases, the capacitance associated with the depleted layer at the poly-Si/gate dielectric interface becomes significant, making it necessary to consider alternative gate electrodes. The search for metallic gates faces many challenges since they must have compatible work functions, thermal/chemical interface stability with underlying dielectric, and high carrier concentration. Metal gates promise to solve several issues, such as poly-gate electrode depletion effects, boron penetration, stability with alternate high-k dielectrics, and decreased gate resistance as devices are scaled down further.

Metal gates are promising candidates to replace the conventional polycrystalline silicon gate electrode. Midgap metal can afford a simpler CMOS processing scheme, since only one mask and one metal would be required for the gate electrode. For alternative metal gate electrode, in addition to the work function requirements, the metal gate and the high-k gate dielectric should be mutually compatible and not interdiffuse or react at the MOSFET thermal budget. Binary metal alloys of Ru and Ta as candidates for CMOS gate electrodes have been proposed. It was reported that Ru-Ta alloys are excellent n-MOSFET gate electrode candidates since they exhibit low work functions and demonstrate superior thermal stability compared to Ta. These metal alloys also offer work function tuning capability. Moreover, by increasing the Ru concentration of this alloy, excellent PMOS gate characteristics were achieved. An intermixed stack of Ru and Ta has been investigated as a route to obtaining ease of integration.

For scaled CMOS devices, a major drawback of midgap metals is that since the band gap of Si is fixed at 1.1 eV, the threshold voltage for any midgap metal on Si will be 0.5 V for both n- and p-MOSFETs. Since voltage supplies are expected to be <1.0 V for sub-130 nm CMOS technology, a V_{th} of 0.5 V is much too large, as it would be difficult to turn on the device. Lowering of the V_{th} would require a lowering of the doping concentration, which would degrade the short-channel characteristics. Therefore, the ideal situation calls for two metals with dual work function: 4 eV for n-MOSFETs and 5 eV for p-MOSFETs. This will enable low-threshold voltages without degraded short-channel effects. For example, the work function value of Al could produce a V_{th} of 0.2 V for NMOS, while the higher work function value of Pt could achieve V_{th} (0.2 V) for p-MOSFETs.

5.1.4 High-k Gate Dielectrics

The successful scaling of silicon-based CMOS technology has been attributed to the prevailing gate dielectric: silicon dioxide. Silicon dioxide exhibits excellent properties, such as remarkable interface quality and robust reliability. However, as silicon dioxide thickness is scaled down, the gate leakage current due to direct tunneling process increases exponentially. For example, for silicon dioxide thinner than 4 nm, every 5 Å reduction in the oxide thickness will result in about two orders of magnitude increase in the direct tunneling current, in which the major challenge is the scaling of the gate dielectric. Traditional gate dielectric, silicon dioxide, has touched its fundamental limit for the 90 nm technology node because the tunneling current increases exponentially as the thickness of the gate dielectric scales down. To continue the scaling trend of the gate dielectric, materials with high permittivity (high-k) have been intensively investigated as possible replacements of silicon dioxide. Several high-k materials have been shown to be promising, such as HfO_2, but many critical integration issues have to be solved for use in MOSFET technologies. As gate oxide thickness decreases, the capacitance associated with the depleted layer at the poly-Si/gate dielectric interface becomes significant, making it necessary to consider alternative gate electrodes. The search for metallic gates also faces many challenges since they must have compatible work functions, thermal/chemical interface stability with the underlying dielectric, and high carrier concentration.

Intel has been at the forefront in addressing the above challenges by successfully driving transistor innovations from the research phase to mainstream CMOS manufacturing. Innovations introduced by Intel to overcome traditional scaling limitations for n-MOSFETs are the following: (1) uniaxial process-induced strain for mobility enhancement starting at the 90 nm CMOS technology node, (2) epitaxial SiGe S/D (e-SiGe), (3) SiN capping layers, (4) high-k gate dielectric introduced at the 45 nm CMOS technology node to replace SiO_2 to reduce gate leakage, and (5) metal gate introduced at the 45 nm CMOS technology node to replace the polysilicon gate to enable oxide thickness scaling. p-MOSFET strain implementation has the following features: (1) SiGe epitaxial S/D formed by Si recess etch and selective strained SiGe epigrowth, (2) strained SiGe-induced large lateral compression in the channel, resulting in higher mobility, (3) SiGe S/D improvement of parasitic resistance by reducing salicide interface resistance, and (4) strained SiGe having a smaller hole barrier height at the silicide interface.

5.2 Multigate Transistors

The multiple-gate field-effect transistor is a promising device architecture for the 45 nm CMOS technology node and beyond. Transition from the planar bulk to the multigate architecture facilitates the target subthreshold performance while still keeping the channel doping concentration low, if ultra-thin Si films and metal gates can be used to control SCEs and adjust the threshold voltage, respectively. Multigate field-effect transistors include double-gate FinFETs, tri-gate FETs, omega-FETs, pi-gate FETs, and gate-all-around FETs, which have been reported to achieve enhanced performance with CMOS-compatible processing. Benefits of multigate FETs include: (1) can harvest 20% more current per chip area, (2) better subthreshold swing due to full depletion, (3) more resistant to random dopant fluctuations, and (4) suppress stress proximity effects. However, these nonplanar devices suffer from a high parasitic resistance due to the narrow width of their source/drain (S/D) regions. The key operation differences and technological issues for each device type are summarised in Table 5.1. Schematics of various types of multigate devices are shown in Figure 5. 3.

Specific features of some of the devices are described below. In gate-all-around (GAA) FETs the gate material surrounds the channel region on all sides. The threshold voltage of GAA FETs is independent of substrate bias due to the complete electrostatic shielding of the channel body. High drive current, excellent gate control revealed by low SCE, and near-ideal subthreshold slope were demonstrated in these devices [3]. The vertical double-gate devices are in their nature modified tri-gate devices. The thick oxide (hard mask) on top of the fin isolates the top gate electrode, and in this way it offers an alternative solution to the "corner effect." The concept of a tri-gate device with sidewalls extending into the buried oxide (called a Π-gate) has been proposed. The gate sidewall extensions effectively act as a back gate through a lateral field effect in the buried oxide. The Π-gate device is simpler to manufacture than the GAA and offers electrical characteristics and short-channel properties close to those of GAA MOSFETs.

TABLE 5.1

Operational Characteristics and Design Considerations for Various Multigate Devices

DG Design	Current Direction	Electric Field from Gate	Design Considerations
Planar	‖ to substrate	⊥ to substrate	Precise control of silicon thickness and bottom gate dimension, gate alignment
Fin	‖ to substrate	‖ to substrate	High aspect ratio/short pitch fin definition, nonplanar gate stack patterning
Vertical	⊥ to substrate	‖ and ⊥ to substrate	Active area hard mask removal and surface prep, layout efficiency
GAA	‖ or ⊥ to substrate	Variable	Access for gate stack deposition and etch, active area dimension uniformity

Source: After Smith, C. E., Advanced Technology for Source Drain Resistance Reduction in Nanoscale FinFETs, PhD thesis, University of North Texas, 2008.

5.3 Double-Gate MOSFET

The double-gate (DG) MOSFET is a promising structure for scaling CMOS into the sub-15 nm gate length regime because of its excellent suppression of short-channel effects (SCEs) for a given equivalent gate oxide thickness. Planar double-gate transistors take advantage of the conventional planar manufacturing processes to create double-gate devices, avoiding more difficult lithography requirements associated with nonplanar, vertical transistor structures. In planar double-gate transistors the channel is positioned between two independently fabricated gate oxide stacks. The addition of the second gate electrode helps to control the potential lines, originating

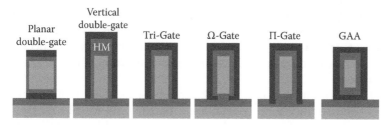

FIGURE 5.3
Fin cross section schematics of the different architectures of multigate devices. (After Shickova, A., Bias Temperature Instability Effects in Devices with Fully-Silicided Gate Stacks, Strained-Si, and Multiple-Gate Architectures, PhD thesis, Katholieke Universiteit Leuven, 2008.)

from the drain bias, and in this way suppresses the short-channel effects. The primary challenge in fabricating such structures is achieving satisfactory self-alignment between the upper and lower gates. Advanced SOI MOSFETs with thin body (thinner than 50% of L_g) thickness (T_{Si}) can suppress the leakage current, which makes this approach technically challenging. In double-gate architecture, the presence of two gates across an ultra-thin body (UTB) helps to reduce the SCE and thus provide a better subthreshold slope. This significantly reduces the subthreshold leakage current for a given I_{on}. For example, the DG device does not need to have high-channel doping to scale because it is defined by body thickness, which is normally 50 to 70% of the gate length to suppress the SCE effectively. As a result, mobility degradation and statistical dopant fluctuation problems can be eliminated. Similarly, X_j is also defined by the body thickness; thus, the shallow junction can be realised relatively easily without developing complicated junction implantation techniques.

Many different methods have been proposed to fabricate DG devices, but most of them suffer from technical challenges, mainly due to the process complexity. For example, the vertical devices with pillar-like channels have a large gate overlap capacitance, and the required processes are very complicated. Devices fabricated on ultra-thin silicon-on-insulator (SOI) wafer (Figure 5.4(a)) are able to achieve a smaller I_{off} by eliminating the leakage path, which is far away from the gate control. When L_g is scaled down to less than 15 nm, according to the International Technology Roadmap for Semiconductors, FinFET or a multigate device structure, as shown in Figure 5.4(b), will be required to control the I_{off} more effectively. As shown in Figure 5.4(c), the drain current can flow from the source to the drain on the top surface and on both sidewall surfaces of the fin. The surface orientations are different, with the top surface being (100) and the sidewall surfaces being (110). The mobility of the carriers traveling in the different surfaces will also be different. For example, it is well known that the hole mobility is higher on a (110) plane than on the (100) [4]. Multiple-gate transistor structures have superior scalability over conventional planar metal-oxide-semiconductor transistor structures, and enable gate length scaling well beyond the 32 nm technology generation. The performance of the multigate device will depend a lot on the dimension of the fin width (W_{fin}). Devices with smaller L_g will usually require smaller W_{fin} for better SCE control. However, the decrease in W_{fin} is accompanied by an increase in series resistance, which degrades the drive current.

FinFET is the most manufacturable double-gate (DG) structure due to process compatibility with conventional planar bulk MOSFETs. However, the channel surface (fin sidewall) roughness induced by photolithography and dry etching degrades carrier mobilities without a subsequent surface smoothening process. Advanced transistor structures such as multigate field-effect transistors improve carrier mobilities further because a heavily doped channel is not necessary to control short-channel effects, compared

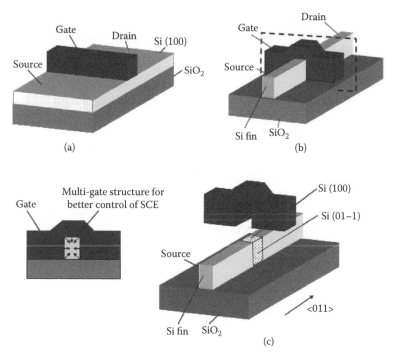

FIGURE 5.4
Schematics of a (a) planar device fabricated on an SOI wafer, (b) FinFET or multiple-gate device, and (c) cross section showing the control by a multigate structure and the different planes of conduction. (After Ming, T. K., Strain Engineering for Advanced Transistor Structure, PhD thesis, National University of Singapore, 2008.)

with the bulk Si MOSFET. Enhancement of multigate performance via process-induced strain has been investigated recently by several groups. The effects of strain on the performance of n-channel and p-channel multigate with {110}/<110> and {100}/<100> surface orientation/current direction have also been addressed.

5.4 Ω-FinFET

Ω-FinFET is known to be the most manufacturable structure due to self-aligned gate electrodes compatible with the conventional planar bulk CMOS process. The Ω-FinFET design, named after the similarity between the Greek letter omega and the shape in which the gate wraps around the fin, is another variation of the tri-gate structure. The fin undercut allows the gate electrodes to extend partially below the fin, and in this way to offer better control of the electric field. The Ω-FinFETs have unique features, such as high heat dissipation to the Si substrate, no floating body effect, and low defect density, while having the key advantages of the silicon-on-insulator (SOI)-based CMOS technology. The Ω-FinFET has a top gate like the conventional UTB-SOI, sidewall gates like FinFETs, and special gate extensions under the silicon body. The Ω-FinFET is basically a field-effect transistor with a gate that almost covers the body. However, the manufacturability of these types of device structures is still an issue. Many methods have been proposed to fabricate these devices, but most of them suffer from technical challenges, mainly due to the process complexity. Aggressively scaled FinFET structures suffer significantly from degraded device performance due to large source/drain series resistance, and to mitigate, several methods such as maximising contact area, silicide engineering, and epitaxially raised S/D have been explored. Strained Si technology is beneficial for enhancing carrier mobilities to boost I_{on}. Both electron and hole mobilities can be improved by applying stress to induce appropriate strain in the channel, e.g., tensile strain for n-MOSFETs and compressive strain for p-MOSFETs. The effect of strain on mobility can be understood by considering the stress-induced changes in the electronic band structures of Si.

The novel device designs require 3D process and device simulations. FinFET is a nonplanar device and is inherently 3D in nature. Therefore, for FinFETs, any meaningful process or device simulation must be performed in three dimensions. Synopsys tools such as SProcess and SDevice address these needs. Figure 5.5 shows process simulation results for 25 nm gate length FinFETs. It shows that a tensile process-induced strain has been evolved in the fin.

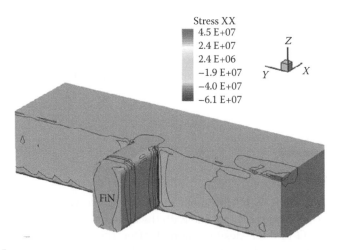

FIGURE 5.5
Stress (ε_{xx}) distributions in channel for Ω-FinFET. (After Maiti, T. K., Process-Induced Stress Engineering in Silicon CMOS Technology, PhD thesis, Jadavpur University, 2009.)

5.5 Tri-Gate FinFET

Tri-gate transistors employ a single gate stacked on top of two vertical gates, allowing for essentially three times the surface area for carriers to travel. In the technical literature, the term *tri-gate* is sometimes used generically to denote any multigate FET with three effective gates or channels. The tri-gate devices having more geometrical dimensions than the planar devices are exposed to a risk from electric field concentration around the fin corners. This corner effect is alleviated by appropriate corner rounding processing. For strained tri-gate FinFETs, contributions from the top gate and the sidewalls should be studied separately because the top gate and sidewalls have different surface orientations, but the underlying physics is the same. Schematic diagrams of a tri-gate single-fin FinFET and a multifin FinFET are shown in Figure 5.6. FinFET devices have more geometrical dimensions than the conventional planar devices. In addition to the gate length, one can also define the fin width (W_{fin}), fin height (H_{fin}), defined by the silicon film thickness, the distance between two adjacent fins (S), and the distance from the gate edge to the source/drain pads, called fin extensions (L_{ext}). Depending on the number of active gate electrodes around the fin, the multigate devices can roughly be classified in groups of double-, triple-, and quadruple-gate devices. A lot of work has been published on the processing, performance, and modelling of multigate devices.

Even though multiple-gate field-effect transistors have several advantages, such as stringent geometric scaling requirements of their planar counterparts, they suffer large parasitic resistance owing to the extremely narrow source drain regions [5]. In case of a FinFET, a large S/D series resistance component is the contact resistance between the semimetallic silicide and the heavily doped semiconducting portion of the silicon fin. From a materials perspective, modifying the properties of the silicide/silicon interface is an attractive option to reduce this contact resistance. This can be achieved either through the use of novel silicides or NiSi alloys, or by altering the silicon dopant density. A list of source drain resistance components, starting from the contact via and migrating toward the channel, is given in Table 5.2.

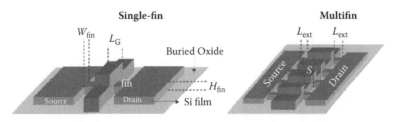

FIGURE 5.6
Schematic diagrams of a tri-gate single-fin and multifin FinFETs. (After Shickova, A., Bias Temperature Instability Effects in Devices with Fully-Silicided Gate Stacks, Strained-Si, and Multiple-Gate Architectures, PhD thesis, Katholieke Universiteit Leuven, 2008.)

TABLE 5.2

S/D Parasitic Resistance Components in a MOSFET

Resistance Component	Symbol	Dependence	Comments
Contact resistance between via and silicide	R_{CON} (via to silicide)	Contact area, barrier height between liner and silicide	Industry standards are Ti/TiN (W plug-Al back end) and Ta/TaN (Cu damascene)
Sheet resistance of the silicide	R_S (silicide)	Silicide resistivity, thickness, and x-y dimensions	Can be dominated by device layout (i.e., gate to drain distance) or 3D structure
Spreading resistance due to transition between S/D pad and 3D silicide over fin	R_{SPsil} (pad to fin)	Contact area and relative volume differences between pad silicide and fin silicide	Only present for fin arrays with S/D pads
Contact resistance between silicide and doped fin extension	R_{CON} silicide to HDD or R_{CON} silicide to ext	Contact area between silicide and silicon fin, barrier height between silicide and doped fin, surface dopant concentration	May be modulated at the interface via dopant or impurity pileup, can exist at multiple junctions (HDD or ext)
Sheet resistance of the HDD fin	R_S HDD fin	Doping concentration, dopant activation efficiency, fin dimensions	Dopant solid solubility-limited conventional-implant issues are reduced in epitaxial S/D portion of fin
Spreading resistance between HDD and fin extension under the spacer	R_{SPsil} HDD to ext	% cross section reduction from S/D to ext under the spacer, dopant concentration gradient	Spread from unsilicided HDD to fin extension = spread from epitaxially thickened HDD to fin extension
Resistance of fin extension	R_{ext}	Dopant concentration and activation	Lateral diffusion profile postactivation anneal determines gate overlap/underlap

Source: After Smith, C. E., Advanced Technology for Source Drain Resistance Reduction in Nanoscale FinFETs, PhD thesis, University of North Texas, 2008.

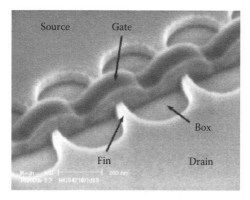

FIGURE 5.7
Tilted SEM image of a typical multifin multigate device with poly-Si gate. (After Shickova, A., Bias Temperature Instability Effects in Devices with Fully-Silicided Gate Stacks, Strained-Si, and Multiple-Gate Architectures, PhD thesis, Katholieke Universiteit Leuven, 2008.)

The contact resistance between the via and silicided S/D region is dominated by the contact area and work function difference between the two metallic conductors. A tilted scanning electron microscope (SEM) image of a typical multifin multigate device is shown in Figure 5.7.

5.6 FinFETs Using Gate-Induced Stress

FinFET is a promising device structure for scaled CMOS logic/memory applications in the 22 nm technology node and beyond. FinFETs employ a very thin undoped body to suppress subsurface leakage paths and hence reduced SCEs. An undoped or lightly doped body eliminates threshold voltage variations due to random dopant fluctuations and enhances carrier transport in the channel region, resulting in a higher ON current. FinFET is an example of a self-aligned double-gate MOSFET built on an SOI substrate and was designed to suppress SCE. A tri-gate MOSFET was developed by Intel composed of multiple gates with a higher surface area for electrons to travel. For FinFET manufacturing, a lot of process challenges need to be addressed due to difficult fin/gate patterning in the 3D structure, conformal doping to the fin, and high access resistance in an extremely thin body. The fin/gate patterning can be improved by optimisation of the patterning stack, patterning scheme, and etch chemistry. FinFET device fabrication has some compatibility with planar CMOS processing techniques. The starting material is a (100) surface-oriented silicon-on-insulator (SOI) wafer. Active area patterning of the SOI material by reactive ion etching (RIE) results in fin structures with (100) top and (110) side surfaces sitting on top of the buried oxide (BOX). A simplified process flow includes the active region, gate, contact, and metal level patterning. The spacer, implant, and silicide process modules are all self-aligned, as in standard planar CMOS processing. Tri-gate FinFET devices have the following advantages: provide more drive current, work in three dimensions, have a geometry advantage, and are fully depleted. A high-performance tri-gate fully depleted CMOS with 60 nm physical gate lengths has been demonstrated that exhibits lower leakage than a standard planar CMOS.

Among the various approaches, introduction of stress by metal-nitride gate is promising. It has been reported that fully silicided metal gate can induce strain in the transistor channel, and the localised strain could be exploited to enhance the performance of aggressively scaled transistors. FinFET structures allow the use of low-channel dopant concentration, and avoid problems associated with random dopant fluctuation. The threshold voltage of FinFETs can be set through gate work function engineering using metal gates, which additionally eliminate the gate depletion effect and dopant penetration problem for improved drive current. For n-channel FinFET devices, the optimal gate work function lies between the midgap and the conduction band of Si, which necessitates the use of metal gates. A simple and cost-effective technique used to incorporate strain in the channel region of FinFET devices has been reported [6]. It is shown that the metal gate can affect the transistor performance through the stress developed during the fabrication process. However, gate work function tuning, process integration, and compatibility with gate dielectric continue to be the major challenges in metal gate technology development. Annealing of a TaN gate electrode capped with a SiN layer leads to the exertion of a compressive stress on the Si fin. This results in a significant enhancement of the drive current in n-channel FinFETs. Mesa-isolated n-channel FinFETs with TaN gates were fabricated on SOI with a (001) surface and 45 nm thick Si.

Figure 5.8 shows the mechanism by which channel stress could be induced by the metal gate. In the strained-channel FinFET the metal gate electrode tends to expand more than the SiN capping layer or the Si fin during the S/D anneal. With the presence of the SiN capping layer, a limited expansion of the TaN gate in the upward direction takes place and results in a compressive stress being exerted onto the Si fin, as illustrated in Figure 5.8(a). This compressive stress in the channel can be retained even after the SiN capping layer is removed. Figure 5.8(b) illustrates that due to the unique structure of the FinFET device, a constrained expansion of the metal gates on both the left and the right side of the fin effectively compresses or squeezes the fin on at least two sides.

The I_d-V_d characteristics of a 75 nm gate length LG FinFET are shown in Figure 5.9(a), with the current being normalised by two times the fin height H_{fin}. The strained-channel FinFET gives a significantly higher drive current

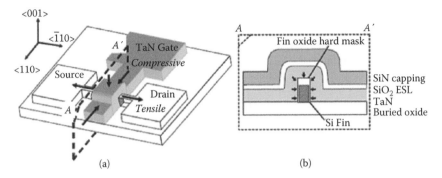

(a)   (b)

FIGURE 5.8
Schematic showing how the TaN gate layer can compressively stress the Si fin channel from three directions. The cross section schematic illustrates the compressive stress exerted perpendicular to the fin body during S/D implant activation anneal. (After Ming, T. K., Strain Engineering for Advanced Transistor Structure, PhD thesis, National University of Singapore, 2008.)

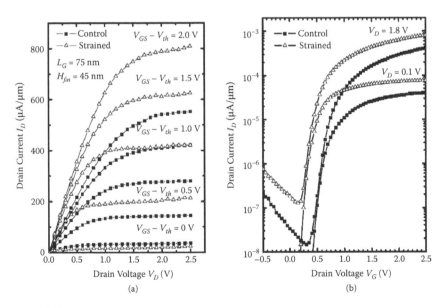

FIGURE 5.9
(a) I_d-V_d characteristics of control and strained FinFET devices at various gate overdrives. The strained FinFET has a significantly higher drive current. (b) Subthreshold characteristics of the control and strained FinFET devices at $V_d = 0.1$ V and $V_d = 1.8$ V. (After Ming, T. K., Strain Engineering for Advanced Transistor Structure, PhD thesis, National University of Singapore, 2008.)

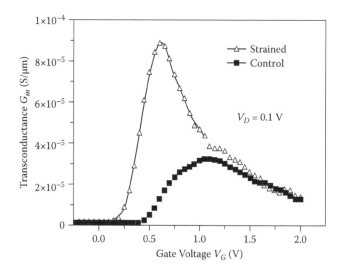

FIGURE 5.10
Comparison of transconductance of the strained and control devices. The higher peak trans-
conductance seen for the strained device indicates a higher mobility. (After Ming, T. K.,
Strain Engineering for Advanced Transistor Structure, PhD thesis, National University of
Singapore, 2008.)

than the control FinFET. The subthreshold swing of the devices is comparable, as shown in Figure 5.9(b). Furthermore, they also demonstrate similar
drain-induced barrier lowering (DIBL). Transconductance measurements, as
plotted in Figure 5.10, show a higher peak linear transconductance for the
strained-channel FinFET compared to the control FinFET, indicating electron mobility enhancement as a result of the strain. It can also be observed
that the strained FinFET exhibits a lower-threshold voltage than the control
FinFET. This could be contributed by the lowering of the conduction band
energy due to the strain effect.

The drain current I_d vs. gate voltage V_g characteristics of a FinFET with a
condensed SiGe S/D and the control device are shown in Figure 5.11(a). The
L_g = 26 nm FinFET with condensed SiGe S/D shows a subthreshold swing of
~100 mV/decade and drain-induced barrier lowering of 0.13 V/V. It can also
be observed that the additional condensation step does not degrade the performance of the FinFET. The difference in DIBL between the control and the
FinFET with condensed SiGe S/D has been attributed to the control device
having a smaller effective length due to process differences.

The I_d-V_d characteristics of the devices are plotted in Figure 5.11(b) at various gate overdrives (V_g-V_{th}). At a gate overdrive of –1.2 V, FinFET with condensed SiGe S/D shows a 28% higher I_{dsat} than the control device. This is
attributed to a recessed Ge profile and an increased Ge concentration for

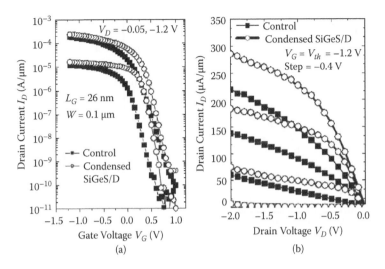

FIGURE 5.11
(a) $I_d - V_g$ characteristics of FinFET devices having an L_g of 26 nm. (b) $I_d - V_d$ characteristics of FinFET devices at various gate overdrives ($V_g - V_{th}$). (After. Ming, T. K., Strain Engineering for Advanced Transistor Structure, PhD thesis, National University of Singapore, 2008.)

larger strain effects. FinFET with condensed SiGe S/D also shows a larger peak transconductance than the control device, as observed in Figure 5.12(a), indicating a higher hole mobility, which can be attributed to the enhanced strain effect. As shown in Figure 5.12(b), the two devices have comparable source/drain series resistances.

5.7 Stress-Engineered FinFETs

The technique of inducing stress by using a tensile (for n-MOSFET) or compressive (for p-MOSFET) SiN_x capping layer is attractive because of its relatively simple process and its extendibility from bulk Si to silicon-on-insulator (SOI) MOSFETs. In this section, the impact of tensile and compressive capping layers on electron and hole mobilities is investigated for Si fins with {100} sidewalls and <100> current flow direction, and Si fins with {110} sidewalls and <110> current flow direction, which are optimal for maximum electron and hole mobilities, respectively [7, 8]. The effects of various structural parameters (gate electrode thickness, gate length, and fin aspect ratio) need to be studied to provide insight for strain engineering in nonplanar FinFET structures.

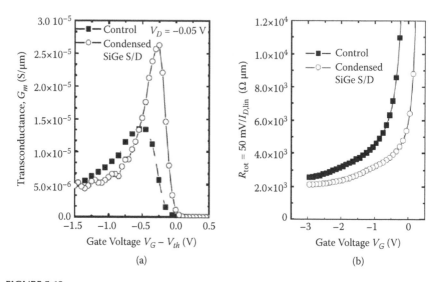

FIGURE 5.12
(a) Comparison of transconductance Gm at the same gate overdrive, illustrating an enhancement of 91% for the FinFET with condensed SiGe S/D over the control device. (b) Extraction of series resistance by examining the asymptotic behavior of the total resistance at large gate bias. (After Ming, T. K., Strain Engineering for Advanced Transistor Structure, PhD thesis, National University of Singapore, 2008.)

The effects of strain on the performance of FinFETs with {110}/<110> and {100}/<100> surface orientation/current direction have been investigated [9]. Substrate-induced strain was studied, and it was found that sSOI improves {110} and {100} electron mobility by 60 and 30%, respectively. Although {110} hole mobility is degraded by 35%, {100} hole mobility is enhanced by up to 18%. Therefore, sSOI is suggested for performance enhancement of {100} CMOS FinFETs, or {110} CMOS FinFETs with selective strain relaxation in p-MOSFETs. A tensile capping layer is expected to provide dramatic enhancement (>100%) in {100} electron mobility, while a compressive capping layer is expected to provide a modest amount (<25%) of {110} hole mobility enhancement. Therefore, dual-stress capping layers with hybrid orientations are suggested as a promising performance booster of CMOS FinFETs. Mobility enhancement is greater for fins with a high aspect ratio (greater than 1), so that greater performance enhancement is expected for double-gate FET (FinFET) vs. tri-gate FinFET devices.

Figure 5.13 shows the 3D structure used for simulations. The 100 nm thick SiNx capping layer has a uniform hydrostatic stress of either 1 GPa (tensile) or –1 GPa (compressive). The bottom surface is the bottom of the 400 nm thick buried oxide. It is assumed that a thin gate oxide layer will have a negligible effect on the stress transfer from the capping layer to the channel, and so it was not included in the simulated structure for

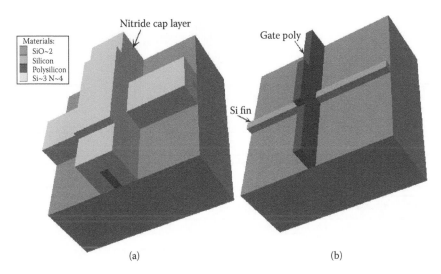

FIGURE 5.13
Three-dimensional structure used for simulations. Nominal values: BOX thickness = 400 nm, fin width = 50 nm, fin height = 50 nm, gate length = 50 nm, fin thickness = 50 nm, fin length = 1 µm, gate poly-thickness = 150 nm, nitride thickness = 100 nm, orientation = (100).

simplicity. A thick dielectric hard mask on top of the fin was not included. It has been observed that the stress profiles are almost identical, except for a change in sign. The amount of induced stress in the channel depends on the distance between the capping layer and the fin, which increases toward the bottom of the fin due to the nonzero thickness of the gate electrode. Thus, the induced stress profile is nonuniform from the top to the bottom of the fin. Figure 5.14 shows the mobility enhancement contours in three directions. Figures 5.15 and 5.16 show the device performance enhancement in the drain current due to stress.

Strain effects on FinFETs have also been studied [10]. The total hole mobility of the FinFET with respect to the stress is shown in Figure 5.17, compared with the single-gate (110)- and (001)-oriented p-type devices at the inversion charge density of $1 \times 10^{13}/cm^2$. In the calculation of the single-gate devices, the doping density is taken to be $1 \times 10^{17}/cm^3$. This is a low doping density compared with the contemporary CMOS technology. Even so, the FinFET shows significantly greater mobility than the bulk devices. If larger doping density is applied, the mobility advantage of the FinFET would be even larger. When 3 GPa uniaxial compressive stress is applied to a FinFET, about 300% enhancement of the mobility is expected, compared to only 200% enhancement for a bulk (110)-oriented transistor, as shown in Figure 5.18. Even though the (001)-oriented p-MOSFET shows greater relative enhancement (over 400%), the absolute mobility is still lower than that of the FinFET due to its low mobility with no stress.

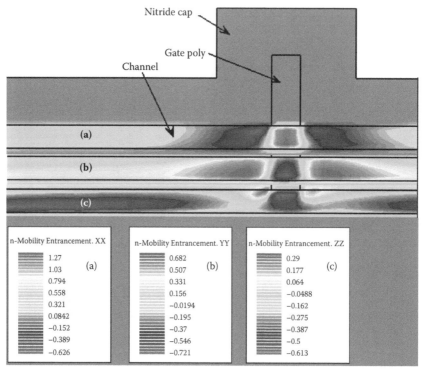

FIGURE 5.14
Three-dimensional mobility enhancement factors along the channel direction.

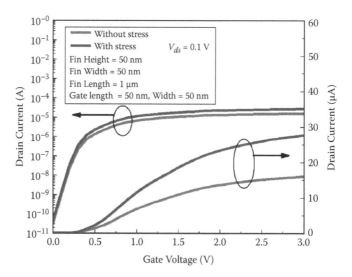

FIGURE 5.15
I_d-V_g characteristics of stress-engineered FinFET devices.

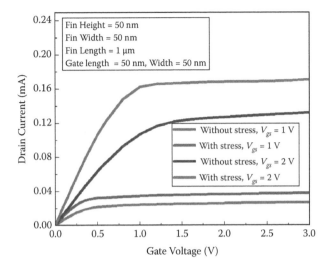

FIGURE 5.16
I_d-V_d characteristics of stress-engineered FinFET devices. Current enhancement due to stress is shown.

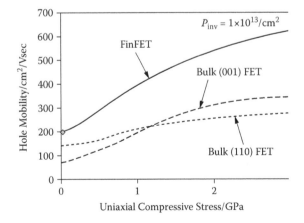

FIGURE 5.17
Hole mobility of FinFETs under uniaxial stress compared with bulk (110)-oriented devices at a charge density of $1 \times 10^{13}/cm^2$. (After Sun, G., Strain Effects on Hole Mobility of Silicon and Germanium p-Type Metal-Oxide-Semiconductor Field-Effect-Transistors, PhD thesis, University of Florida, 2007.)

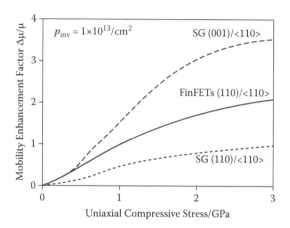

FIGURE 5.18
Hole mobility enhancement factor of FinFETs under uniaxial compressive stress at charge density of $1 \times 10^{13}/cm^2$. (After Sun, G., Strain Effects on Hole Mobility of Silicon and Germanium p-Type Metal-Oxide-Semiconductor Field-Effect-Transistors, PhD thesis, University of Florida, 2007.)

5.8 Layout Dependence

The layout-dependent stress effect is first observed and reported from shallow trench isolation (STI) stress [11]. The insertion and placement of active areas are optimised by STI stress to improve the circuit performance [12]. The stress profile is sensitive to the primary layout parameters, such as channel length and source/drain diffusion length. STI width effect has been investigated and shown to enhance circuit performance. A stress-aware layout design has been proposed to reduce leakage power. Layout-dependent stress effects are also being observed in the state-of-the-art strain technologies. The stress dependence provides circuit designers another alternative to optimise the circuit performance. Thus, interaction between layout and circuit performance needs to be accurately predicted using stress models. Toward this, the traditional efforts resort to TCAD simulation to extract the stress level from the entire layout and analyse performance enhancement. Compact models that capture the dependence on primary layout parameters, temperature, and other device characteristics, such as mobility, velocity, and threshold voltage in state-of-the-art strain technologies like e-SiGe and DSL stress techniques, have been reported [13].

The main techniques to introduce uniaxial stress include embedded SiGe technology (e-SiGe), dual-stress liner (DSL), stress memorisation technique (SMT), and the parasitic stress from shallow trench isolation (STI). Embedded SiGe technology embeds SiGe in the source and drain area to introduce compressive stress for p-MOSFETs. The amount of performance enhancement depends on both the applied stress magnitude and circuit layout parameters, such as gate length, source/drain size, and the distance from gate edge to STI because of the nature of mechanical stress in silicon: nonuniform distribution. DSL introduces the stress by depositing a highly stressed silicon nitride layer, tensile stress for the n-MOSFET region, and compressive stress for the p-MOSFET region, over the entire wafer to elevate carrier mobility. In SMT, the stress in the channel is transferred from the stressed deposited dielectric and is memorised during the recrystallisation of the active area and poly-gate when thermal annealing is activated. STI stress results from the difference in thermal expansion coefficients between SiO_2 and Si. It is an intrinsic stress source and not intentionally built up for enhancing device performance enhancement. Process-induced stress has been effectively applied for the 90 nm node and beyond. Since the stress is nonuniformly distributed in the channel, the enhancement in carrier mobility, velocity, and threshold voltage shift strongly depend on circuit layout, leading to systematic performance variations among transistors. However, special layout engineering is required for practical application because process-induced strain is localised strain and depends on the physical dimensions of the transistor, such as gate length and channel width [14, 15], as well as the surrounding structures. It is important to understand the degree to which the distributions of device parameter (for example, gate length) values of neighbouring, near-neighbouring, and well-separated devices are related to each other (as a function of physical separation).

Layout dependences for stress-enhanced MOSFETs, including contact positioning, the second neighbouring polyeffect, and bent diffusion modeled in 45 nm CMOS logic technology, have been reported [16]. It has been shown that stress effect might be more serious for p-MOSFETs than for n-MOSFETs. Figure 5.19 illustrates the TCAD simulation of stress distribution in a 45 nm standard cell under restrictive design rules, where SiGe with 25% Ge composition is embedded in the S/D area. The stress level is widely different across the cell, depending on transistor size and layout pitch. Such nonuniformity results in pronounced variations among transistors as well as circuit performance, and further increases the complexity in modelling and simulation. The stress effect is weakened by shrinking the pitch of gates. The stress modulation of embedded SiGe depends on the effective area size of SiGe, and therefore the space of a gate and the shape of diffusion influence the performance of MOSFETs. Additionally, the well proximity effect and round shape of patterns in lithography affect the MOSFET performance. For example, the strain-induced threshold voltage shift is mainly dominated by the bottom stress level in the channel, while the entire channel stresses are required to be taken into consideration for the enhancement of mobility. These effects cause the variability of devices in the cells and circuits with random patterns as the feature size is scaled down to 45 nm CMOS logic technology.

Joshi et al. [17] have reported that the trend of the STI stress effect is attributed to the nonuniform stress distribution in the channel, but it does not quantitatively explain how this distribution impacts the electrical properties.

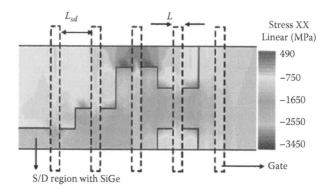

FIGURE 5.19

Top views of stress contours in a five-finger layout pattern with SiGe embedded in the source/drain area. (After Wang, C.-C., Predictive Modelling for Extremely Scaled CMOS and Post Silicon Devices, PhD thesis, Arizona State University, 2011.)

The layout-dependent stress effects are also observed in the state-of-the-art strain technologies. In Figure 5.20(a), TCAD simulation using e-SiGe technology shows the stress profile in the channel, with a higher stress level at the edges and lower stress in the centre of the channel. The stress profile is sensitive to the primary layout parameters, such as channel length and source/drain diffusion length. To capture this layout dependence, Dunga et al. [18] propose a modelling approach to finding an equivalent stress level in the channel accounting for the mobility enhancement, with an assumption that the mobility enhancement is proportional to the applied stress. In Figure 5.20(b), TCAD simulation shows the obvious difference of the equivalent stresses between the shifts in threshold voltage and mobility for different devices with various channel lengths.

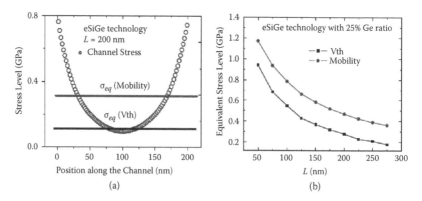

FIGURE 5.20

(a) The stress distribution in the channel. (b) The equivalent stress levels for strain-induced shifts of V_{th} and mobility. (After Wang, C.-C., Predictive Modelling for Extremely Scaled CMOS and Post Silicon Devices, PhD thesis, Arizona State University, 2011.)

5.9 Summary

Process integration issues such as power consumption, leakage current, metal gate electrodes, and high-k gate dielectrics have been discussed. Different types of multigate devices in terms of their architectures were reviewed. The tri-gate devices have been the focus of this chapter, because they are a good compromise between processing complexity and electrical performance. Although the GAA and the ∏-gate structures show better electrical properties, they require more complex and costly processing to be implemented. Layout dependence of strain-enhanced MOSFETs and interaction between layout and circuit performance have been discussed. The stress dependence provides circuit designers an alternative to optimise the circuit performance via stress-aware layout design.

Review Questions

1. Uniaxial process-induced strain for mobility enhancement started at the 90 nm CMOS technology node. (True/False)
2. Epitaxial SiGe S/D (e-SiGe) is used for p-MOSFET performance enhancement. (True/False)
3. SiN capping layers are used for p-MOSFET performance enhancement. (True/False)
4. What are the advantages of high-k gate dielectrics?
5. What are the advantages of metal gate electrodes?
6. What are the disadvantages of polysilicon gate?
7. Strained SiGe induced large lateral compression in channels, resulting in higher mobility. (True/False).
8. SiGe S/D improves parasitic resistance by reducing salicide interface resistance. (True/False)
9. Double-gate MOSFETs have lower short-channel effects. (True/False)
10. Tri-gate FinFETs are at the core of the 22 nm technology node. (True/False)

References

1. International Technology Roadmap for Semiconductors (ITRS). http://public.itrs.net/.
2. H. Iwai, CMOS Scaling toward sub-10nm Regime, *Proc. Int. Symp. Electron Device Microwave Optoelectronic Appl.*, 30–34, 2003.

3. J.-P. Colinge, Ed., *FinFETs and Other Multi-Gate Transistors*, Springer Science + Business Media, LLC, New York, USA, 2008.

4. L. Chang, M. Ieong, and M. Yang, CMOS Circuit Performance Enhancement by Surface Orientation Optimization, *IEEE Trans. Electron Dev.*, 51, 1621–1627, 2004.

5. A. Dixit, A. Kottantharayil, N. Collaert, M. Goodwin, M. Jurczak, and K. De Meyer, Analysis of the Parasitic S/D Resistance in Multiple-Gate FETs, *IEEE Trans. Electron Dev.*, 52, 1132–1140, 2005.

6. T. K. Ming, Strain Engineering for Advanced Transistor Structure, PhD thesis, National University of Singapore, 2008.

7. H. Irie, K. Kita, K. Kyuno, and A. Toriumi, In-Plane Mobility Anisotropy and Universality under Uni-Axial Strains in n and p-MOS Inversion Layers on (100), [110], and (111) Si, *IEEE IEDM Tech. Dig.*, 225–228, 2004.

8. T. Sato, Y. Takeishi, H. Hara, and Y. Okamoto, Mobility Anisotropy of Electrons in Inversion Layers on Oxidized Silicon Surfaces, *Phys. Rev. B*, 4, 1950–1960, 1971.

9. K. Shin, C. O. Chui, and T.-J. King, Dual Stress Capping Layer Enhancement Study for Hybrid Orientation finFET CMOS Technology, *IEEE IEDM Tech. Dig.*, 988–991, 2005.

10. G. Sun, Strain Effects on Hole Mobility of Silicon and Germanium p-Type Metal-Oxide-Semiconductor Field-Effect-Transistors, PhD thesis, University of Florida, 2007.

11. K. Ota, K. Sugihara, H. Sayama, T. Uchida, H. Oda, T. Eimori, H. Morimoto, and Y. Inoue, Novel Locally Strained Channel Technique for High Performance 55nm CMOS, *IEEE IEDM Tech. Dig.*, 27–30, 2002.

12. V. Moroz, G. Eneman, P. Verheyen, F. Nouri, L. Washington, L. Smith, M. Jurczakl, D. Pramanik, and X. Xu, The Impact of Layout on Stress Enhanced Transistor Performance, *Proc. SISPAD*, 143–146, 2005.

13. C.-C. Wang, Predictive Modelling for Extremely Scaled CMOS and Post Silicon Devices, PhD thesis, Arizona State University, 2011.

14. F. Nouri, P. Verheyen, L. Washington, V. Moroz, I. De Wolf, M. Kawaguchi, S. Biesemans, R. Schreutelkamp, Y. Kim, M. Shen, X. Xu, R. Rooyackers, M. Jurczak, G. Eneman, K. De Meyer, L. Smith, D. Pramanik, H. Forstner, S. Thirupapuliyur, and G. S. Higashi, A Systematic Study of Trade-Offs in Engineering a Locally Strained pMOSFET, *IEEE IEDM Tech. Dig.*, 1055–1058, 2004.

15. S. Eneman, P. Verheyen, R. Rooyackers, F. Nouri, L. Washington, R. Degraeve, B. Kaczer, V. Moroz, A. De Keersgieter, R. Schreutelkamp, M. Kawaguchi, Y. Kim, A. Samoilov, L. Smith, P. P. Absil, K. De Meyer, M. Jurczak, and S. Biesemans, Layout Impact on the Performance of a Locally Strained PMOSFET, *Proc. Symp. VLSI Technol.*, 22–23, 2005.

16. E. Morifuji, H. Aikawa, H. Yoshimura, A. Sakata, M. Ohta, M. Iwai, and F. Matsuoka, Layout Dependence Modelling for 45-nm CMOS with Stress-Enhanced Technique, *IEEE Trans. Electron Dev.*, 56, 1991–1998, 2009.

17. V. Joshi, B. Cline, D. Sylvester, D. Blaauw, and K. Agarwal, Leakage Power Reduction Using Stress-Enhanced Layouts, *Proc. DAC*, 912–917, 2008.

18. M. V. Dunga, C.-H. Lin, X. Xi, D. D. Lu, A. M. Niknejad, and C. Hu, Modelling Advanced FET Technology in a Compact Model, *IEEE Trans. Electron Dev.*, 53, 1971–1978, 2006.

6

Noise in Strain-Engineered Devices

C. Mukherjee

Indian Institute of Technology, Kharagpur

The planar device architecture of conventional metal-oxide-semiconductor field-effect transistors (MOSFETs) is limited to scaling beyond 15 nm gate length due to transistor switching criteria. Complementary metal-oxide-semiconductor (CMOS) technology has been scaled during the past 30 years with a drive to continuously increase the density of devices on a chip and increase the switching performance of transistors, the major components of electronic circuits. Toward the end of the ITRS road map [1], in which the channel length is predicted to be aggressively scaled, careful device design consideration is required due to trade-offs between device current drive, short-channel effects, and power consumption. The on-state current (I_{on}) of a MOSFET is represented by

$$I_{on}/W = Q_s(V_{DD})v(V_{DD}) \approx C_G(V_{DD} - V_{th})v(V_{DD}) \tag{6.1}$$

where W is the device's width, V_{DD} is the power supply voltage, V_{th} is threshold voltage, Q_s is the inversion charge density, and v is the velocity near the source region (injection velocity). The power consumption, P_{diss}, can be approximated by [2]

$$P_{diss} = P_D + P_S = \alpha f C_L V_{DD}^2 + V_{DD}\left(I_{leak} + I_{th}10^{\frac{V_{th}}{S}}\right) \tag{6.2}$$

where P_D, P_S, α, f, C_L, and S are dynamic power dissipation, static power dissipation, activity factor, operating frequency, load capacitance, and subthreshold slope, respectively, and I_{leak} represents the total leakage current from gate and junction sources, and I_{th} is the drain current at V_{th}. In order to maintain low power consumption and lower V_{DD} and leakage current, higher V_{th} and steeper S are required according to Equation (6.2). On the other hand, large gate capacitance, low V_{th}, and high velocity are required to achieve a high performance in terms of I_{on}. In addition to the trade-offs for V_{th} and V_{DD}, the choice of high C_G requires a thinner dielectric, which can increase direct tunneling, which enhances the leakage and increases the power consumption. From the electrostatics point of view, high substrate doping is required for aggressively scaled planar devices to control the short-channel effects. The high doping results in increased junction and gate-induced drain lowering (GIDL) degraded on current due to the increased Coulombic scattering and increased variation in threshold voltage. In addition, extension and halo implants needed to control short-channel effects increase source/drain parasitic series resistance, which degrades the current drive. Considering the trade-off between the current drive, short-channel effects, and power consumption, conventional Si MOSFETs fail to satisfy the device requirements that call for new materials and device architectures for future CMOS generations. To enhance the current drive, new channel materials such as strained Si, SiC, SiGe, Ge, and III-V have been extensively investigated over the past 20 years. Uniaxially strained Si technology with tensile liner and embedded SiGe stressors was incorporated into mainstream CMOS production starting at the 90 nm technology node. To further continue scaling and improve the current drive high-permittivity-dielectric (high-κ)/metal gate technology has also been commercialised by Intel in the 45 nm technology node. This has been shown to dramatically improve the gate leakage and power consumption for both n- and p-MOSFET devices. Ultra-thin-body and multigate SOI devices have been shown to provide excellent scalability and immunity to short-channel effects. The geometry enables excellent electrostatic control by the gate, and the lightly doped Si channel dramatically reduces the random dopant fluctuation and V_{th} variation. In addition, these device architectures benefit from lower capacitive parasitic and junction leakage due to the presence of a thick buried oxide. Among various options for multigate device architecture, such as double-gate, tri-gate, etc., the nanowire (NW) channel with a wraparound gate, so called gate-all-around (GAA), has the largest advantage in terms of electrostatic integrity. However, several undesired effects become prominent from the miniaturisation of the device dimensions. One such unwanted effect is a strong increase of the low-frequency noise generated in the transistor as the size of the device decreases. Moreover, there are many unexplored issues regarding the introduction of new materials in complementary metal-oxide-semiconductor (CMOS) technology. Therefore, electrical evaluations of devices using new materials and architectures are highly desired.

Noise is a fundamental problem in science and engineering, recognised for a variety of fields such as telecommunication, nanoelectronics, and biological systems. The noise cannot be completely eliminated, and with small signal strength, the accuracy and measurements are limited in electronic circuits. The low-frequency noise, or $1/f$ noise, is the excess noise at low frequencies whose power spectral density (PSD) approximately depends inversely on the frequency, and therefore escalates at low frequencies. The $1/f$ noise originating from the transistors is a severe obstacle in analogue circuits. The $1/f$ noise is, for example, upconverted to undesired phase noise in voltage-controlled oscillator (VCO) circuits, which can limit the information capacity of communication systems [3]. Phase noise is a difficult problem in wireless transceivers and radio frequency (RF) circuits; RF oscillators are designed with sensitive phase noise requirements. Both the unwanted signal from an adjacent channel and the desired single are downconverted by the oscillator and are mixed at the output. The phase noise from the downconverted interfering signal overlaps with the desired signal, corrupting the signal-to-noise ratio. The detrimental effects of the phase noise can be limited by placing the channels farther apart in frequency, at the cost of reduced information capacity of the communicating system. $1/f$ noise in semiconductor devices poses a significant problem for VCOs when upconverted to undesired phase noise at small frequency offsets from the carrier frequency, and therefore set the ultimate separation limit of two channels [3, 4]. Figure 6.1 illustrates the phase noise spectrum and its different physical origins in RF circuits. The main drawback of oscillators implemented in CMOS technology compared to bipolar technology is the inferior $1/f$ noise characteristics in the CMOS circuits, thus limiting its use in high-performance oscillators [4]. This makes the study and understanding of $1/f$ noise mechanisms in oscillators so important for reducing the phase noise originating from device $1/f$ noise by proper circuit design. In oscillator circuits, frequency of oscillation is a function of the device current. Low-frequency noise in the current is directly translated to low-frequency noise in the frequency of oscillation, and in turn to phase noise. According to Hajimiri and Lee [4], noise at frequencies near integer multiples of the oscillation frequency contributes significantly to the total phase noise.

Apart from $1/f$ noise problems in RF circuits, the downscaling of the device dimensions reflects a downscaling of the voltage levels too, which lowers the signal-to-noise ratio. In effect, the $1/f$ noise may soon become a major concern not only in analogue circuits, but also in digital ones [6]. The relative

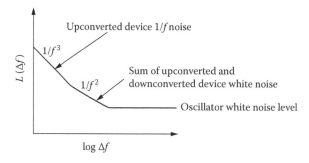

FIGURE 6.1
Schematic illustration of the phase noise spectrum. (After Haartman, M. V., Low-Frequency Noise Characterization, Evaluation, and Modelling of Advanced Si- and SiGe-Based CMOS Transistors, PhD thesis, Royal Institute of Technology (KTH), Sweden, 2006.)

noise level already is a problem in RF and analogue applications, and soon exceeds the limits for a reliable device operation also in digital applications. Overcoming the $1/f$ noise in electronic circuits and devices is an extremely important challenge for the future. Low-frequency noise measurements are also an important tool for device diagnostics. The $1/f$ noise is very sensitive to trap and defects in the device, and is strongly related to physical processes such as trapping and release phenomena, electron scattering mechanisms, and phonon processes. The low-frequency noise can therefore be used as the information-carrying signal to evaluate and get insight into the physics and properties of a particular system, and estimate the quality and reliability of a device [7, 8].

In order to minimise the device $1/f$ noise, an understanding of the noise mechanisms, the underlying physics, and the location of the sources is necessary. Still today, after several decades of debate, the exact origin of the $1/f$ noise is, in many aspects, unclear [9]. In this chapter, the $1/f$ noise sources and their origins have been discussed, including the physical properties of mobility fluctuation noise, one of the most debated proposed $1/f$ noise mechanisms. An elaborate analysis and modelling of the $1/f$ noise in terms of carrier number fluctuations, mobility fluctuations, substrate voltage effects, gate voltage dependency, stress, and correlated mobility fluctuations is presented in terms of the device physics and the properties of current transport. For extracting information from traps, the time domain of random telegraph signal (RTS) noise is also explored. The low-frequency noise study in various emerging devices presented here is intended for use in designing nanoscale devices with new materials and architectures for optimising the $1/f$ noise performance.

6.1 Noise Mechanisms

Currents and voltages in an electronic circuit show random fluctuations (thereby causing a sharp rise and fall) around their DC bias values due to fluctuations in the physical processes that govern the electronic carrier transport. The desired signal is difficult to detect distinctly if the background noise power is significantly high compared to the signal strength. Noise is a fundamental problem in science and engineering since it cannot be eliminated. In devices with highly scaled dimensions, the accuracy of measurements is thus limited by setting a lower limit on signal strength that can be accurately detected and processed. The importance of noise characterisation has been acknowledged in a variety of fields, such as telecommunication, nanoelectronics, mesoscopic structures, and biological systems. However, noise not only poses a problem that should be avoided as much as possible, but it also can actually be used as a tool to evaluate and get insight into the properties and reliability of a particular system.

Characterisation of the low-frequency noise in electronic devices gives important information of the device physics and reliability, such as scattering processes, traps, and defects.

6.2 Fundamental Noise Sources

The total output current, $I(t)$, of a device can be written as the sum of bias current (I_{bias}) and the randomly fluctuating noise current ($i_n(t)$) as $I(t) = I_{bias} + i_n(t)$. The external sources that cause fluctuations in the current are not considered in this case. There are some fundamental physical processes, which act as the sources of noise, that can generate the random fluctuations in the current (or voltage) in a device. These noise sources are discussed below and described in terms of the power spectral density (PSD) of the noise current.

6.2.1 Thermal Noise

Thermal noise (also known as Nyquist, Johnson, diffusion, velocity fluctuation, or white noise) originates from the random thermal movement of electrons, and is present at all frequencies with a flat frequency response (PSD). The phenomenon of thermal noise can be thought of as the thermal excitation of the carriers in a resistor. Due to scattering the velocity of the electron changes randomly. At a particular time instant, there could be more electrons moving in a certain direction than electrons moving in the other directions in a random manner, resulting in a small net current. This current fluctuates randomly in strength and direction, with the average over (long) time being zero. If a piece of material with resistance R and temperature T is considered, the PSD of the thermal noise current is found to be

$$S_I(f) = 4kT/R \quad \text{or} \quad S_V(f) = 4kTR \tag{6.3}$$

where k is Boltzmann's constant. The thermal noise exists in every physical resistor and resistive part of a device and sets a lower limit on the noise in an electric circuit. In bipolar transistors, thermal noise can be modeled as originated from base resistance and the collector impedance. In field-effect transistors, the existence of thermal noise comes from the physical channel resistance between the drain and gate (when the device is on and conducting current).

6.2.2 Shot Noise

The current flowing across a potential barrier, like the p-n junction, fluctuates due to the random movement of the electronic charge (electrons). The current across a barrier is given by the number of carriers, each carrying the charge q, flowing through the barrier during a period of time. A shot noise current is generated at low frequencies when the electrons cross the barrier independently and randomly. At higher frequencies it transforms to white noise. The shot noise current fluctuates with a PSD of [10]

$$S_I = 2qI \qquad (6.4)$$

The physics behind shot noise is closely related to the thermal noise phenomenon. A p-n junction has a nonlinear resistance; the spectral density of the noise current is half the thermal noise for the dynamic resistance associated with the p-n junction. The reason behind the factor 1/2 is basically that the current is essentially flowing in one direction across the p-n junction. In bipolar transistors, the sources of shot noise are located at the depletion region of each junction. The recombination (at the base-emitter junction) of minority carriers generated at the base contributes to the shot noise, whereas in the collector-base junction, the minority carriers generated at the emitter and base contribute to the shot noise. Shot noise in FETs, on the other hand, is attributed to the gate leakage current.

6.2.3 Generation–Recombination Noise

Generation–recombination (g-r) noise in semiconductors originates from random capture or emission of carriers by localised charge centres (or traps), thereby causing random fluctuation in the carrier number. If carriers are trapped at some critical spots, the trapped charge can also induce fluctuations in the mobility, diffusion coefficient, electric field, barrier height, space charge region width, etc. Localised defect states within the forbidden band gap are referred to as traps, the physical origin of which are due to the presence of various defects or impurities in the semiconductor bulk or at the surface. In MOSFETs the inversion charge may be trapped or de-trapped in these defect states, causing current or voltage level to fluctuate. The carrier transitions in a semiconductor mainly consist of generation of an electron/hole pair, recombination of a free electron and hole, and trapping of electrons and holes in empty traps.

A trap may be neutral or charged in its empty state (depending on whether it is a donor trap or acceptor trap). From the Langevin differential equation governing how the number of carriers N depend on time,

$$\frac{d\Delta N}{dt} = -\frac{\Delta N}{\tau} + H(t) \qquad (6.5)$$

where $H(t)$ is a random noise term, ΔN is the fluctuation in the number of carriers, and τ is the time constant. The PSD of the carrier fluctuation can be derived [11]:

$$S_N(f) = \frac{4\overline{\Delta N^2}\tau}{[1+(2\pi f\tau)^2]} \qquad (6.6)$$

Here, f is the frequency. The shape of the spectrum given by Equation (6.6) is called Lorentzian. G-r noise is only significant when the Fermi level is within a few kT in energy of the trap energy level. In this case, the capture time τ_c and the emission time τ_e are almost equal. If the Fermi level is far above or below the trap level, the trap will be filled or empty most of the time, and very few transitions would occur to produce noise. The current density in n-type bulk semiconductor can be written as

$$J = \sigma E = (qn\mu_n + qp\mu_p)E \approx nq\mu_n E \tag{6.7}$$

If n fluctuates ($n = N/V$, where V is the volume), the current density fluctuates as

$$S_J(f) = \frac{S_N(f)}{N^2} J^2 \tag{6.8}$$

Thus, S_J decreases with increasing N as $1/N^2$. The variation of the PSD with the number of carriers is one way to distinguish noise originating from traps from noise related to fluctuations in the mobility.

6.2.4 Random Telegraph Signal (RTS) Noise

A special case of g-r noise is RTS noise, also known as burst noise or popcorn noise, which is displayed as random switching events in the time-domain voltage or current signal, as shown in Figure 6.2. In a MOSFET, if a carrier is

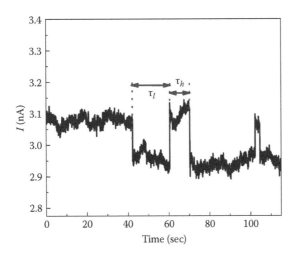

FIGURE 6.2
A typical RTS showing the low and high current levels in the time-domain signal.

trapped in a single trap or localised defect state, the current or voltage signal displays a random shift in the level denoting a change in the channel resistance. In bipolar transistors, however, the trapping/de-trapping process has different mechanisms, which will be discussed later in the chapter, involving the tunneling of carriers across the p-n junction potential barrier. The two-level RTS signal signifies only one active trap. However, when multiple traps are involved, the current (or voltage) can switch between two or more states resembling a RTS waveform due to random trapping and de-trapping of carriers, and the phenomenon is much more difficult to explain in order to identify the trapping/de-trapping process. For simple two-level RTS pulses with equal height ΔI and Poisson distributed mean time durations in the lower state τ_l and in the higher state τ_h, the PSD of the current fluctuations is derived as [12]

$$S_I(f) = \frac{4(\Delta I)^2}{(\overline{\tau_l} + \overline{\tau_h})\left[\left(1/\overline{\tau_l} + 1/\overline{\tau_h}\right)^2 + (2\pi f)^2\right]} \tag{6.9}$$

Mainly two types of traps are identified depending on the nature of trapping mechanism. They are donor and acceptor traps. The donor trap is charged when it emits an electron (i.e., empty) and is neutral when it captures. The acceptor trap is, contrary to the donor trap, charged when it captures an electron and neutral when empty. In MOSFETs, the channel resistance increases with the charged trap state, changing the current (or voltage) to a high state. Clearly, the donor trap causes high current level after emission of carriers, and the acceptor trap causes the high current level when it captures an electron.

Depending on the values of the mean time constants of the RTS, the traps can be characterised by two types. These are the slow traps, with high values of time constants, and the fast traps, with the time constants being very small (order of a few 0.1 ms). The PSDs of the RTS noise and the g-r noise are both of the Lorentzian type. G-r noise can be modeled as a sum of RTSs from one or more traps with identical time constants, and it is a RTS in the time domain only if a small number of traps are involved. RTS noise is an interesting phenomenon from device physics point of view since the random switching process due to a single trap can be studied in the time domain. It is a well-accepted theory that RTS is caused by a single carrier controlling the flow of a large number of carriers, rather than a large number of carriers being involved in the trapping/de-trapping process [8]. From RTS noise characterisations, interesting information about the trap energy, capture and emission kinetics, and spatial location of the traps inside the semiconductor device can be acquired. The multilevel RTS is due to the activation of multiple traps near the quasi-Fermi level. With smaller area devices, only single traps are active, as the number of traps is fewer and the RTS becomes a simple two-level signal, with a Lorentzian PSD ($1/f^2$).

6.2.5 1/f Noise

1/f noise, also called flicker noise or pink noise, is the low-frequency noise with fluctuations with a PSD proportional to $1/f^\gamma$, with γ close to 1, usually in the range 0.7–1.3. The PSD for 1/f noise takes the general form

$$S_I(f) = \frac{KI^\beta}{f^\gamma} \tag{6.10}$$

where K is a constant and β is a current exponent. There are so many theories regarding the 1/f noise mechanisms, of which most prominent are the carrier number fluctuation (surface phenomenon) and mobility fluctuation (bulk phenomenon) theories. The mechanisms behind flicker noise are still a long-debated topic and an interesting research topic. The generally accepted origins of the 1/f noise are attributed to conductivity fluctuations, damage in crystal structures, and traps due to defects in semiconductors. 1/f fluctuations in the conductance have been observed in the low-frequency part of the spectrum (10^{-6} to 10^6 Hz) in most conducting materials and a wide variety of semiconductor devices [11, 13]. From Equation (6.7) also, it is clear that there are essentially two physical mechanisms behind any fluctuations in the current: fluctuations in the mobility or fluctuations in the number of carriers (g-r noise). G-r noise from a large number of traps can produce 1/f noise if the time constants of the traps are distributed as [14]

$$g(\tau) = 1/\ln(\tau_2/\tau_1)\tau, \tau_2 < \tau < \tau_1$$

$$g(\tau) = 0, otherwise \tag{6.11}$$

The factor $\ln(\tau_2/\tau_1)$ is for normalisation purposes. The total noise PSD ($S_{tot}(f)$) from superposition of the g-r noise from many traps distributed according to $g(\tau)$ yields, combining Equations (6.6) and (6.11),

$$S_{tot}(f) = \int_0^\infty g(\tau)S_{g-r}(\tau)d\tau = \frac{1}{\ln(\tau_2/\tau_1)} \int_{\tau_1}^{\tau_2} \frac{1}{\tau} \frac{K\tau}{1+(2\pi f\tau)^2} d\tau \tag{6.12}$$

$$or, S_{tot}(f) \approx \frac{K}{4\ln(\tau_2/\tau_1)f}, for 1/2\pi\tau_2 \ll f \ll 1/2\pi\tau_1$$

An illustration is shown in Figure 6.3, where g-r noise from four individual traps with different time constants adds up to a $1/f^\gamma$ spectrum, with γ approximately 1. Some remarks are necessary about the addition of g-r noise spectra. The superposition of g-r noise in producing a 1/f spectrum assumes

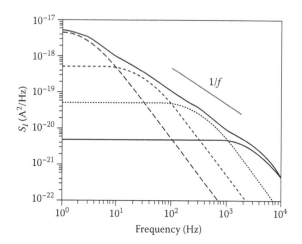

FIGURE 6.3
Superposition of four Lorentzians that gives a total spectrum approximately showing $1/f$ dependence over several decades of frequency. (After Haartman, M. V., Low-Frequency Noise Characterization, Evaluation, and Modelling of Advanced Si- and SiGe-Based CMOS Transistors, PhD thesis, Royal Institute of Technology (KTH), Sweden, 2006.)

that the g-r noise from the traps can simply be added with the traps being isolated from each other, and they do not interact. G-r noise is obtained with a time constant given by the reciprocal sum of all time constants if interaction occurs [15]. Also, the traps are assumed to couple in the same way to the output current so that the value of K is the same for all traps.

The second well-accepted mechanism behind $1/f$ noise is known as mobility fluctuations. It was first described by Hooge with the following empirical formula for the resistance fluctuations (S_R) [16]:

$$\frac{S_R(f)}{R^2} = \frac{\alpha_H}{fN} \tag{6.13}$$

The dimensionless parameter α_H, known as Hooge's parameter, was first suggested to be constant and equal to 2×10^{-3}. Later, it was found that α_H depends on the crystal quality, and the value may change in different materials. Only phonon scattering contributes to the mobility fluctuations, and the factor $1/N$ results from independent mobility fluctuations by each of the N conducting carriers. The conductivity σ is given as

$$\sigma = \frac{q}{V}\sum_{i=1}^{N}\mu_i = qN\overline{\mu_i}/V = qn\overline{\mu_i} \tag{6.14}$$

The conductivity fluctuates due to fluctuations in the individual carrier mobilities μ as

$$\Delta\sigma = \frac{q}{V}\sum_{i=1}^{N}\Delta\mu_i \Rightarrow \overline{\Delta\sigma^2} = \frac{q^2}{V^2}\sum_{i=1}^{N}\overline{(\Delta\mu_i)^2} = \frac{q^2}{V^2}N\overline{(\Delta\mu_i)^2} \qquad (6.15)$$

Using Equation (6.13), the noise power spectral density is given as

$$\frac{S_\sigma}{\sigma^2} = \frac{S_\mu}{\mu^2} = \frac{1}{N}\frac{S_{\mu_i}}{\mu_i^2} = \frac{S_R}{R^2}, with \frac{S_{\mu_i}}{\mu_i^2} = \frac{\alpha_H}{f} \qquad (6.16)$$

which means that α_H is proportional to the variance of the relative mobility fluctuation for each carrier, independent of the number of carriers. The mobility fluctuation noise is always present, and $1/f$ noise in metals and bulk semiconductors is dominated by mobility fluctuations [13]. In MOS transistors, the conducting channel near the surface under the gate oxide also contributes to noise, with traps in the gate oxide as the dominant $1/f$ noise source. However, the mobility fluctuation noise model explains the $1/f$ noise in p-MOSFETs better [9].

Another theory on $1/f$ noise mechanism that needs mentioning is the quantum noise theory proposed by Handel [17]. In this theory, the $1/f$ noise is explained by electron scattering due to infrared photon emission. When electrons are scattered they lose momentum, causing emission of photons with energy $h\nu$, which depends on the frequency ν. This leads to a probability of photon emission proportional to $1/f$ giving the $1/f$ noise fluctuations in the scattering cross section. There are, however, many flaws in this theory from practical and theoretical viewpoints. The originally proposed model by Handel was confirmed by Van Vliet's [18] quantum electrodynamical theory, but many of Handel's later additions were rejected. The Hooge's parameter described by this model for silicon has a value of about 10^{-8}, which deviates far from the range of values from 10^{-6} to 10^{-3} for conventional Si MOSFETs. Although, quantum $1/f$ noise theory sets the lower limit for $1/f$ noise, clearly other sources are more dominant in the majority of devices.

The latest addition to mobility fluctuation noise theory, proposed by Musha and Tacano, suggests that an energy partition among weakly coupled harmonic oscillators in an equilibrium system is subjected to $1/f$ fluctuations [19]. Jindal and van der Ziel [20] suggested that the phonon population also demonstrates g-r fluctuations that may cause phonon scattering, and in effect lead to mobility fluctuation and electrical g-r noise. Mihaila proposed that an inelastic tunneling process with active phonon vibrations may be the origin of both the number and mobility fluctuation noise [21].

6.3 1/*f* Noise in MOSFETs

The origin of the 1/*f* noise in MOS transistors has been much debated concerning whether carrier number fluctuation noise due to traps in the gate oxide or bulk mobility fluctuations dominates the 1/*f* noise. In 1957, McWorther presented a 1/*f* noise model based on quantum mechanical tunneling transitions of electrons between traps in the gate oxide and the channel [22]. The tunneling time varies exponentially with distance from the trap, and the 1/*f* noise is obtained for a trap density that is uniform in both energy and distance from the channel interface. The McWorther model is widely accepted for simplicity and excellent agreement with experiments, especially for n-MOSFETs. However, the mobility fluctuation noise model explains the 1/*f* noise in p-MOSFETs better [9]. It was later explained by the unified flicker noise theory that a trapped carrier also affects the surface mobility through Coulombic interaction. This correlated mobility fluctuation model gave a correction to the number fluctuation noise model, which resolves the deviations of the theory for p-MOSFETs. However, the correction factor was criticised for being unfeasibly high since screening was not accounted for. Also, the carrier mobility at the surface is reduced compared to the bulk mobility due to additional surface scattering (by acoustic phonons and surface roughness), which has an impact on the mobility fluctuations. Moreover, the Hooge mobility noise is sensitive to the crystalline quality, which is deteriorated close to the interface. The most feasible explanation for the higher 1/*f* noise, with the carriers being in close proximity of the gate oxide surface, is increased mobility fluctuation noise.

6.3.1 Number Fluctuations

The physical mechanism behind the number fluctuation noise is interaction between slow traps in the gate oxide and the carriers in the channel, which is schematically illustrated in Figure 6.4. The interaction and exchange of carriers between the channel and the oxide traps results in a fluctuation in the surface potential, thereby causing variation in inversion charge density and effectively noise in the drain current. Although the fluctuation in inversion charge density causes no current flow, the drain current is needed to sense the fluctuation externally. The fluctuating oxide charge density δQ_{ox} is equivalent to a variation in the flat-band voltage (V_{fb}).

$$\delta V_{fb} = -\delta Q_{ox}/C_{ox} \tag{6.17}$$

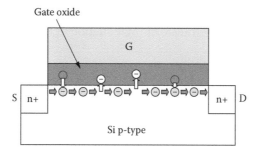

Gate oxide

G

S n+   n+ D

Si p-type

FIGURE 6.4
Schematic illustration of electrons in the channel of a MOSFET moving in and out of traps, giving rise to fluctuations in the inversion charge density, and thereby the drain current. (After Haartman, M. V., Low-Frequency Noise Characterization, Evaluation, and Modelling of Advanced Si- and SiGe-Based CMOS Transistors, PhD thesis, Royal Institute of Technology (KTH), Sweden, 2006.)

The fluctuation in the drain current $I_D = f(V_{fb}, \mu_{eff})$ then yields [23]

$$\delta I_D = \frac{\partial I_D}{\partial V_{fb}} \delta V_{fb} + \frac{\partial I_D}{\partial \mu_{eff}} \frac{\partial \mu_{eff}}{\partial Q_{ox}} \delta Q_{ox}$$

$$or, \delta I_D = -g_m \delta V_{fb} + \frac{I_D}{\mu_{eff}} \frac{\partial \mu_{eff}}{\partial Q_{ox}} \delta Q_{ox} \tag{6.18}$$

With $\alpha = \frac{1}{\mu_{eff}^2} \frac{\partial \mu_{eff}}{\partial Q_{ox}}$ (for n-MOSFET),

$$\delta I_D = -g_m \delta V_{fb} - I_D \mu_{eff} \alpha C_{ox} \delta V_{fb} \tag{6.19}$$

So, transferring Equation (6.19) into frequency domain, current noise power spectral density becomes

$$S_{I_D} = S_{V_{fb}} \left(1 + \frac{\alpha \mu_{eff} C_{ox} I_D}{g_m} \right)^2 g_m^2 \tag{6.20}$$

The first term in the parentheses in Equation (6.20) is due to the fluctuating number of inversion carriers, and the second term to mobility fluctuations correlated to the number fluctuations. Note that α can be negative or positive depending on if the mobility increases or decreases upon trapping a charge according to Equation (6.18). The power spectral density of the flat-band voltage fluctuations is calculated by summing the contributions from all traps in the gate oxide by transforming Equation (6.17) [24]:

$$S_{Q_{ox}} = S_{V_{fb}} C_{ox}^2 = \frac{q^2}{W^2 L^2} \int_{E_V}^{E_C} \int_0^W \int_0^L \int_0^{t_{ox}} 4 N_t f(E)(1 - f(E)) \frac{\tau}{1 + (2\pi f \tau)^2} dx \, dy \, dz \, dE \quad (6.21)$$

where $f(E) = 1/[1 + e^{(E-E_{fn,p})/kT}]$ is the Fermi function, and N_t is the density of traps in the gate dielectrics at the quasi-Fermi level (in $cm^{-3} eV^{-1}$). Only these traps contribute to the $1/f$ noise, with the other traps being permanently filled or empty. In the McWorther model it is assumed that trapping and de-trapping occur through tunneling processes; the trapping time constant is given as

$$\tau = \tau_0(E)e^{z/\lambda}, with, \lambda = \left[\frac{4\pi}{h} \sqrt{2m^* \Phi_B} \right]^{-1} \quad (6.22)$$

for tunneling from the interface to the trap located at position z in the gate oxide. The tunneling attenuation length λ is predicted by the Wentzel–Kramers–Brillouin (WKB) theory, Φ_B is the tunneling barrier height seen by the carriers at the interface, and m^* is the effective mass of channel carriers in the gate oxide. The time constant τ_0 is often assumed as 10^{-10} s, and $\lambda \approx 1$ Å for the Si/SiO$_2$ system. This yields $z = 2.6$ and 0.7 nm for frequencies of 0.01 Hz and 1 MHz, respectively. Thus, oxide traps located too close to the channel interface are too fast to give $1/f$ noise, and those located more than ~3 nm from the interface are too slow to contribute. By inserting Equation (6.22), the integral in Equation (6.21) can be evaluated as

$$S_{V_{fb}} = \frac{q^2 kT \lambda N_t}{f^\gamma WLC_{ox}^2} \quad (6.23)$$

The frequency exponent γ deviates from 1 if the trap density is not uniform in depth. γ is less than 1 when the trap density near the oxide/semiconductor interface is higher than inside the trap density inside the gate oxide, and γ is greater than 1 for the opposite case.

The simulated bias dependence of the normalised drain current noise PSD, S_{ID}/I_D^2, in the number fluctuation model with drain currents ranging from subthreshold to strong inversion regimes using Equation (6.20), $\alpha = 0$, and a constant arbitrary N_t, is shown in Figure 6.5. S_{ID}/I_D^2 varies approximately as $1/(V_{GS} - V_T)^2 \propto 1/Q_i^2$ in strong inversion. S_{ID}/I_D^2 decreases more rapidly with drain current as g_m is reduced at high-gate-voltage overdrives. In the subthreshold region, on the other hand, S_{ID}/I_D^2 is almost constant since $g_m = I_D q/mkT$. The physical explanation is that change in

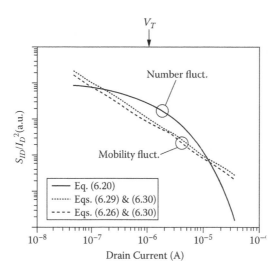

FIGURE 6.5
Simulation of the drain current noise using both the number and the mobility fluctuation noise models, with $\alpha = 0$ and constant N_t for the number fluctuations, and with constant α_H for the mobility fluctuations. (After Haartman, M. V., Low-Frequency Noise Characterization, Evaluation, and Modelling of Advanced Si- and SiGe-Based CMOS Transistors, PhD thesis, Royal Institute of Technology (KTH), Sweden, 2006.)

oxide charge density is greater than the change in inversion charge density as the charge trapped in the oxide is not only supplied from the inversion charge but also from the depletion and interface trap charges. The normalised drain current noise in the subthreshold region can be written as [25]

$$\frac{S_{I_D}}{I_D^2} = \frac{q^4 \lambda N_t}{kTf^\gamma WL(C_{ox} + C_d + C_{it})^2} \tag{6.24}$$

where C_{ox}, C_d, and C_{it} are the oxide, depletion capacitance, and capacitance due to interface trapped charge.

A trap density that increases toward the conduction or valence band edges results in a faster fall-off of S_{I_D}/I_D^2 than $1/Q_i^2$ due to the band bending, resulting in a faster change in the oxide trap energy compared to the interface traps. In this case γ is greater than 1 and increases with gate bias. Studies on RTS noise in MOSFETs show that thermally activated phonon-assisted capture and emission of carriers play an important role [8]. If the time constant of the trap is written as

$$\tau_{th} = \tau_{0,th} e^{E/kT} \tag{6.25}$$

$1/f$ noise is obtained for an even distribution of traps in energy. The problem with this theory is the difficulty to find a physical process with the property given by Equation (6.25). The emission time for a thermally activated trap depends exponentially on the activation energy, but the capture time is normally independent of energy.

6.3.2 Mobility Fluctuations

The drain current noise power spectral density is given by Hooge's empirical formula due to channel carrier mobility fluctuations:

$$\frac{S_{I_D}}{I_D^2} = \frac{q\alpha_H}{fWLQ_i} \tag{6.26}$$

which is derived from Equation (6.13) with the number of carriers N in the channel replaced by WLQ_i/q. In the linear region, $Q_i = C_{ox}(V_{GS} - V_T)$, and thus the normalised drain current noise depends inversely on the gate voltage overdrive. Typical values for α_H range between 10^{-3} and 10^{-6}. Values about 10^{-7} have also been observed for buried channel Si p-MOSFETs [26]. The mobility $1/f$ noise is suggested to be generated by phonon scattering [27]. Different scattering mechanisms responsible for the channel carrier mobility fluctuation depend on the effective electric field and the inversion charge density in different ways. So, α_H not only depends on the semiconductor materials or the technology, but is also governed by the bias conditions. Each scattering process, j, generates mobility fluctuation noise, with the Hooge's parameter of that process being $\alpha_{H,j}$. If all the scattering processes are independent of each another, Matthiessen's rule can be applied to sum them up for calculating the effective mobility μ_{eff}.

$$\frac{1}{\mu_{eff}} = \sum_j \frac{1}{\mu_j} \tag{6.27}$$

Power spectral density is

$$\frac{S_{I_D}}{I_D^2} = \frac{S_{\mu_{eff}}}{\mu_{eff}^2} = \sum_j \left(\frac{\mu_{eff}}{\mu_j}\right)^2 \frac{q\alpha_{H,j}}{fWLQ_i}, with, \alpha_H = \sum_j \left(\frac{\mu_{eff}}{\mu_j}\right)^2 \alpha_{H,j} \tag{6.28}$$

The relation in Equation (6.26) is only valid for a uniform carrier density. In the saturation region, the carrier density varies parabolically along the channel and reaches zero at the drain. The total channel drain current noise can be evaluated by summing up the noise contribution from each channel segment from source to drain as

$$\frac{S_{I_D}}{I_D^2} = \frac{q\alpha_H}{fWL^2}\int_0^L \frac{dx}{Q_i(x)} = \frac{q\alpha_H}{fWL^2}\int_0^{V_{DS}} \frac{W\mu_{eff}}{I_D}dV = \frac{q\alpha_H\mu_{eff}V_{DS}}{fL^2I_D} \qquad (6.29)$$

This equation is valid for all regions of operation, but V_{DS} is replaced with $V_{DS,sat}$ for $V_{DS} > V_{DS,sat} = (V_{GS} - V_T)/m$ (with m being the ideality factor). However, the drain current and total charge density Q_i is independent of V_{DS} when $V_{DS} \gg mkT/q$. The mobility $1/f$ noise is also independent of V_{DS} [28]:

$$\frac{S_{I_D}}{I_D^2} = \frac{\alpha_H\mu_{eff}\,2kT}{fL^2I_D} \qquad (6.30)$$

The mobility in subthreshold is not easily characterised; the value can be estimated from the mobility value close to the threshold voltage. From Figure 6.5, it can be deduced that the number fluctuation noise only becomes dominant near the threshold, whereas the mobility fluctuation noise is prominent both at the subthreshold region and the strong inversion region.

6.4 Noise Characterisation in MOSFETs

Noise measurement is a very sensitive and complex task as the signal strength is very small, the lower limit being ~1 pA. DC bias current and disturbances from other electronic equipment add to the difficulty of measurements. The measurement setup must be designed carefully with appropriate shielding from external noise and using batteries as power sources to avoid disturbances to be injected in the circuits. Noise measurements are done in the frequency domain with the PSD measured by the dynamic signal analyser performing the fast Fourier transform on the time-domain signal. Time-domain analysis of RTS is also a valuable noise characterisation tool.

6.4.1 Noise Measurements as a Diagnostic Tool

Low-frequency noise measurements can be used as a valuable tool for quality and reliability analysis and lifetime assessment of electronic devices. The noise measurements reveal the noise mechanisms as well as the location of noise sources inside the semiconductor devices studied from bias and geometry dependence. For example, if the drain current noise power spectral density is caused due to source/drain resistance, the noise is independent of the gate length, while if the noise is caused by channel carrier fluctuation, the noise changes with gate length. The noise mechanism can also be revealed from the bias dependence of the low-frequency noise. By varying the gate voltage in a MOSFET the inversion carrier density will change, which reflects on the active noise mechanism. From this bias dependency, the dominant source of the $1/f$ noise, be it mobility fluctuation or number fluctuation noise, can be identified by analysing the resemblance with Equation (6.20) or (6.26). Unlike the straightforward way it is described, the practical measurements pose a lot of problems in identifying the actual mechanisms involved. For example, both the number fluctuation and the mobility fluctuation sources can contribute to the excess noise with comparable magnitude. Practically, there are usually large deviations from the simplistic theoretical description of the mechanisms; the trap density may vary with energy, correlated mobility fluctuations may have a gate voltage dependence similar to that of the Hooge noise, and Hooge's parameter may vary with inversion carrier density and electric field involving some complex mechanisms. The trap density and Hooge's parameter can be used as figures of merit for a given technology or material system. Correlating the noise level to other device parameters such as oxide charge density, interface state density, carrier mobility (especially phonon or Coulomb scattering limited mobility), oxide thickness, etc., can help to establish the noise origin. The basic understanding of the nature of noise sources allows one to interpret the noise data obtained from the noise measurements, and thereby enables one to develop an in-depth knowledge of the semiconductor device physics, including current conduction mechanisms, defects, interface quality, etc. The noise spectroscopy study is essential for not only inspecting the defects and determining the nature and locations of dominant noise source responsible, but also providing an insight on the remedies of the detrimental effects of noise in small geometry devices. For example, if oxide traps are found to be the dominant noise sources behind the $1/f$ noise, reducing the trap density by an improved gate oxidation process will reduce the noise. If mobility fluctuation is revealed to be dominant, the quality of the surface or the interface and the crystal structure will be improving, or incorporation of a strained channel (to improve carrier mobility) may contribute to the reduction of $1/f$ noise. For both mechanisms, a buried channel always provides improved noise performance. The noise due to source/drain resistance can be mitigated by reducing the source/drain resistances, avoiding current crowding, and improving the quality of contacts.

G-r noise, RTS noise, and number fluctuation $1/f$ noise in the MOSFET drain current originate from oxide traps. G-r and RTS noise are only significant when the trap energies are close to the Fermi level energy, and are therefore sensitive to bias and temperature. As RTS noise is caused due to a single trap being active, it is only observed in small area devices or devices with a low background noise. RTS noise can be observed over the mobility $1/f$ noise in MOSFETs with small gate area (usually below 1 μm²) if the following criterion on the number of carriers (N) in the channel is fulfilled [29]:

$$N < 1/4\pi\alpha_H \tag{6.31}$$

With increasing N (increasing gate voltage overdrive), the visibility of RTS over the mobility $1/f$ noise becomes very small. On the other hand, when the $1/f'$ noise and the RTS noise are originated from the same oxide traps, the occurrence of RTS noise depends on gate area but not on bias. The number of traps that can generate $1/f'$ noise can be estimated according to

$$\text{Number of traps} = 4\,kTWLN_t z \tag{6.32}$$

where z is the tunneling distance of a carrier from the gate oxide/channel interface, maximum ~3 nm, N_t is the trap density, and $4\,kT$ is the energy around the Fermi level where the traps are distributed. RTS noise can be observed if the number of active traps is very small. The relative drain current amplitude is related to the trap position inside the oxide z_t as [30]

$$\frac{\Delta I_D}{I_D} = \frac{q}{WL}\left[\frac{g_m}{I_D}\frac{1 - z_t/t_{ox}}{C_{ox}} + \alpha\mu_{eff}\right] \tag{6.33}$$

The trap depth can also be extracted from the variation of the characteristic time constants with gate voltage [31]. For a two-level RTS, considering electrons as the charge carriers and assuming a acceptor trap, the high current level time τ_+ is the time when the trap is filled (emission time τ_e) and the low current level time τ_- is the time when the trap is empty (capture time τ_c) Statistical measurement of τ_+ and τ_- reveals two exponential distributions. The RTS exhibits a Lorentzian spectrum in the frequency domain with a corner frequency in excellent agreement with the harmonic mean of the two transition times.

From the principle of detailed balance, one can write the ratio of the mean emission time τ_e and mean capture time τ_c as [31]

$$\frac{\tau_c}{\tau_e} = g\exp\left(\frac{E_T - E_F}{kT}\right) \tag{6.34}$$

where g is a degeneracy factor usually considered as 1, $(E_T - E_F)$ is the energy level of the trap relative to the Fermi level, T is the absolute temperature, and K is the Boltzmann constant.

In an n-MOSFET, when the gate bias is increased, the trap occupancy is expected to increase, and hence the τ_c/τ_e ratio should change. This change would indicate capture or emission of an electron, and hence provide an insight into the type of trap. With the high current level (charged trap state) identified as the occupied state of the trap, the traps can be identified as acceptor traps. Thus the high current time was considered as τ_e and the low current time as τ_c. Equation (6.34) can be written as [31]

$$\ln\left(\frac{\tau_c}{\tau_e}\right) = -\frac{1}{kT}\left[\begin{array}{c} (E_{Cox} - E_T) - (E_c - E_F) - \phi_0 \\ +q\psi_s + \frac{qz_t}{t_{ox}}(V_{GS} - V_{FB} - \psi_s) \end{array}\right] \tag{6.35}$$

where E_{Cox} is the conduction band edge of the oxide, E_C is the conduction band edge of the silicon, Φ_0 is the difference between the electron affinities of Si and SiO$_2$, t_{ox} is the oxide thickness, V_{GS} is the gate bias, V_{FB} is the flat-band voltage, ψ_s is the surface potential, and z_t is the position of the trap measured from the Si/SiO$_2$ interface. Differentiating in terms of the gate bias, we can obtain the position of traps as

$$\frac{d}{dV_{GS}}\ln\left(\frac{\tau_c}{\tau_e}\right) = -\frac{1}{kT}\left[\frac{qz_t}{t_{ox}}\right] \tag{6.36}$$

The capture and emission times, τ_c and τ_e, are in general governed by Shockley–Read–Hall statistics [32]:

$$\tau_c = \frac{1}{nv_{TH}\sigma_n}, \tau_e = \frac{\exp\left[\frac{(E_T - E_F)}{kT}\right]}{N_C v_{TH}\sigma_n} \tag{6.37}$$

where n is electron density in the vicinity of traps, v_{TH} is the thermal velocity of electrons, σ_n is the electron capture cross section of the traps, N_C is the density of state in the conduction band, and E_T and E_F are the trap energy level and Fermi potential, respectively.

While the energy level and spatial location of the trap can be determined from analysis of the RTS noise, the distribution of traps vs. energy and oxide depth is characterised from the frequency and bias dependence of the number fluctuation $1/f^t$ noise. The trap density as a function of gate bias, which can be related to the Fermi level, can be evaluated by using Equation (6.20). However, one must be cautious with this kind of analysis; the bias dependence could stem from a completely different mechanism.

6.5 Strain Effects on Noise in MOSFETs

6.5.1 n-MOSFET under Tensile Stress

Results of drain current $1/f$ noise measured on an n-MOSFET with a channel length $L = 2$ µm, a width $W = 50$ µm, and a threshold voltage $V_{th} = 0.36$ V under tensile stress is presented in this section. The MOSFET was biased

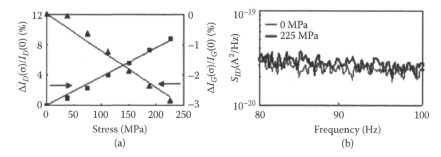

FIGURE 6.6

n-MOSFET under tensile stress: (a) relative changes in drain and gate tunneling currents and (b) comparison of drain current noise PSD for different stresses. (After Lim, J.-S., Strain Effects on Silicon CMOS Transistors: Threshold Voltage, Gate Tunneling Current, and 1/f Noise Characteristics, PhD thesis, University of Florida, 2007.)

in the linear region at $V_G = 0.6$ V and $V_D = 0.1$ V, and six uniaxial tensile stresses were applied up to 225 MPa [33]. The drain (I_D) and gate currents (I_G) are observed to consistently increase and decrease, respectively, as shown in Figure 6.6(a), with increasing tensile stress. Under a stress of 200 MPa, the drain current increases by about 7%, and the gate current decreases by about 2.4%. The drain current noise PSD is observed to increase with tensile stress, as shown in Figure 6.6(b).

6.5.2 n-MOSFET under Compressive Stress

An n-MOSFET with $L = 2$ µm, $W = 50$ µm, and a threshold voltage $V_{th} = 0.28$ V, under six uniaxial compressive stresses up to 189 MPa and biased in the linear region at $V_G = 0.6$ V and $V_D = 0.07$ V, shows a different trend of PSD than the tensile stress case. With increasing compressive stress, the drain current (I_D) decreases and the gate current (I_G) increases, as shown in Figure 6.7(a) [34]. The drain current decreases by about 6%, and the gate tunneling current increases by about 2.5% at a compressive stress level of 200 MPa. The drain current noise PSD is shown to decrease in Figure 6.7(b).

6.5.3 p-MOSFET under Compressive Stress

The 1/f drain current noise power spectral densities are shown in Figure 6.8 for a p-MOSFET with $L = 1$ µm, $W = 50$ µm, and a threshold voltage $V_{th} = -0.36$ V. The results are shown for the device under compressive stresses up to 189 MPa at two different gate biases, $V_G = -0.6$ V and -0.8 V, at the same drain bias, $V_D = -0.1$ V. The drain (I_D) and gate (I_G) currents are commonly observed to increase and decrease, respectively, with increasing compressive stress. The change is about 10% for drain currents and -2% for gate currents

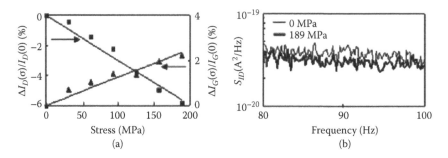

FIGURE 6.7
n-MOSFET under compressive stress: (a) relative changes in drain and gate tunneling currents and (b) comparison of drain current noise PSDs for different stresses. (After Lim, J.-S., Strain Effects on Silicon CMOS Transistors: Threshold Voltage, Gate Tunneling Current, and 1/f Noise Characteristics, PhD thesis, University of Florida, 2007.)

at a compressive stress level of 200 MPa. The noise PSDs increase for both of the two measurements, as shown in Figure 6.8(b).

To obtain more accurate PSD data (as there are high uncertainties in PSD measurements), a more averages are taken. More averaging is required to differentiate even a few percent of change in the noise PSD. Since the drain current 1/f noise PSD generally follows the following frequency dependence,

$$S_{I_D}(f) = \frac{S_{I_D}(1Hz)}{f^\gamma} \tag{6.38}$$

the noise magnitude and exponent can be extracted by a least-squares fit (LSF) of the noise spectrum on a log-log plot. Over the frequency range

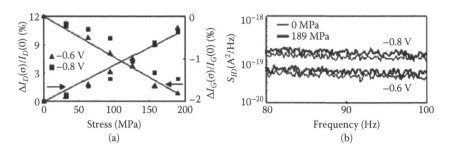

FIGURE 6.8
p-MOSFET under compressive stress: (a) relative changes in drain and gate tunneling currents and (b) comparison of drain current noise PSDs for different stresses. (After Lim, J.-S., Strain Effects on Silicon CMOS Transistors: Threshold Voltage, Gate Tunneling Current, and 1/f Noise Characteristics, PhD thesis, University of Florida, 2007.)

that Equation (6.38) holds, the PSD can be expressed as a linear function of frequency on a log-log plot,

$$\log[S_{I_D}(f)] = \log[S_{I_D}(1Hz)] - \gamma \log[f] \qquad (6.39)$$

Although the PSD data follow $1/f^\gamma$ dependence over a wide frequency range, Equation (6.38) applies only for a smaller range of frequencies locally, and the noise exponent, γ, and the magnitude, S_{I_D} (1 Hz), are frequency independent. But both the noise exponent and the magnitude are affected by applied stress. This is shown for a p-MOSFET PSD in Figure 6.9(a) and (b). For each applied uniaxial longitudinal compressive stress, γ and S_{I_D} (1 Hz) are extracted via LSF of the PSD over a frequency range of 30 Hz to 1 kHz and plotted in Figure 6.9(a) and (b) as a function of stress for a gate bias of –0.6 V. The normalised change in S_{I_D} (1 Hz) relative to the unstressed case,

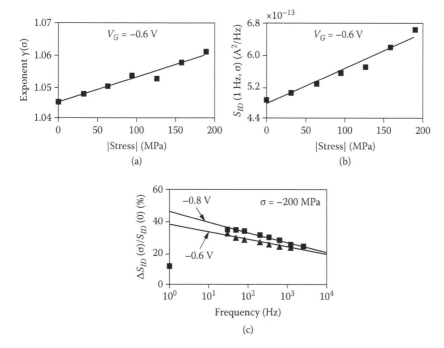

FIGURE 6.9
Analysis of p-MOSFET data under compressive stress: (a) extracted exponent γ value vs. applied stress, (b) extracted noise magnitude vs. applied stress, and (c) relative changes in noise PSD vs. frequency at different gate voltages. The lines are plotted based on the extracted noise magnitude and exponent values, and the symbols are averaged values obtained using 50 pairs of neighbouring noise PSD data. (After Lim, J.-S., Strain Effects on Silicon CMOS Transistors: Threshold Voltage, Gate Tunneling Current, and 1/f Noise Characteristics, PhD thesis, University of Florida, 2007.)

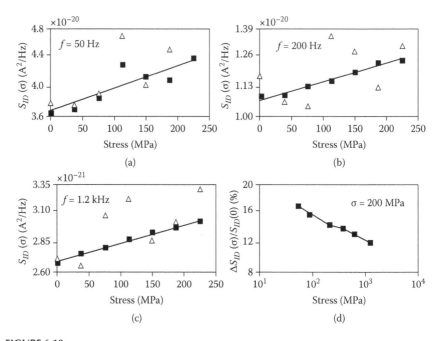

FIGURE 6.10
n-MOSFET noise PSD under tensile stress: (a–c) Changes in noise PSD vs. applied stress at different frequencies. (d) Relative change in noise PSD vs. frequency at a stress of 200 MPa. (After Lim, J.-S., Strain Effects on Silicon CMOS Transistors: Threshold Voltage, Gate Tunneling Current, and 1/f Noise Characteristics, PhD thesis, University of Florida, 2007.)

S_{ID} (1 Hz; σ)/S_{ID} (1 Hz; 0), is shown as a function of stress in Figure 6.9(c). When the PSD follows $1/f^t$ dependence over a wide frequency range, the global and local LSF both show good agreement for two gate biases.

Figures 6.10 and 6.11 show the extracted average drain current noise PSD, $S_{ID}(f; σ)$, at specific frequencies for n-MOSFETs as a function of applied tensile and compressive stress, and the normalised change in PSD, $\Delta S_{ID}(f; σ)/S_{ID}(f; 0)$, is shown as a function of frequency for a specific stress.

Both 1/f noise magnitude and exponent are functions of applied mechanical stress. Therefore, the strain-induced relative change in 1/f noise PSD, referenced at zero stress, can also be expressed from Equation (6.38) as

$$\ln\left(1 + \frac{\Delta S_{I_D}(f;σ)}{S_{I_D}(f;0)}\right) = \ln\left(1 + \frac{\Delta S_{I_D}(1Hz;σ)}{S_{I_D}(1Hz;0)}\right) - \Delta\gamma(σ)\ln[f] \qquad (6.40)$$

where $\Delta S_{I_D}(f;σ) = S_{I_D}(f;σ) - S_{I_D}(f;0)$ and $\Delta\gamma(σ) = \gamma(σ) - \gamma(0)$. Linear relationships of $S_{I_D}(f;σ)$ and $\gamma(σ)$ can be assumed since the applied stresses are small (<250 MPa). The main focus of the study of strain effects on noise PSD

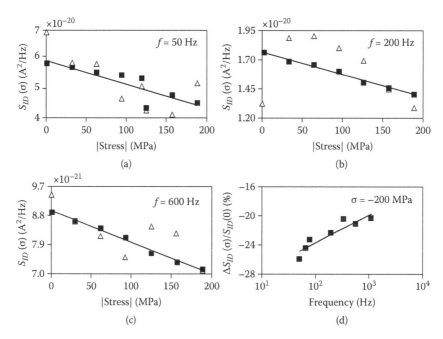

FIGURE 6.11
n-MOSFET noise PSD under compressive stress: (a–c) Changes in noise PSD vs. applied stress at different frequencies. (d) Relative change in noise PSD vs. frequency at a stress of –200 MPa. (After Lim, J.-S., Strain Effects on Silicon CMOS Transistors: Threshold Voltage, Gate Tunneling Current, and 1/f Noise Characteristics, PhD thesis, University of Florida, 2007.)

is on $\Delta S_{ID}(f;\sigma)$ and $\Delta\gamma(\sigma)$, as they vary independently with applied stress. A theoretical model for the stress dependence of drain current noise PSDs is shown in the following section.

6.5.4 Number Fluctuation Model under Strain

6.5.4.1 Mechanisms for Change in Noise PSD under Strain

A number fluctuation model is considered to explain the strain dependence of $1/f$ noise in MOSFETs. As described in Section 6.4.1, the number fluctuation model explains the origin of $1/f$ noise as trapping and de-trapping of channel charge carriers by oxide traps [24]. In ultra-thin gate oxide MOSFETs, even a relatively low gate bias can cause significant band bending in the Si channel, thus causing the Fermi level to lie above the conduction band edge or below the valence band edge. Due to the band bending, the $1/f$ noise PSD mostly results from trapping/de-trapping of channel carriers at oxide traps existing at an energy level above the Si conduction band edge or below the Si valence band edge. In ultra-thin gate oxides, therefore, trapping by two-step or multiphonon processes can be neglected compared to the trapping

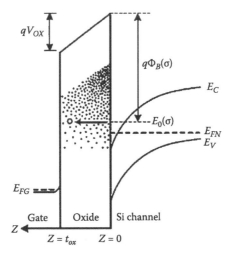

FIGURE 6.12
Schematic band diagram of an n-MOSFET under mechanical stress depicting trapping of channel charge carriers through an elastic direct tunneling mechanism. (After Lim, J.-S., Strain Effects on Silicon CMOS Transistors: Threshold Voltage, Gate Tunneling Current, and 1/f Noise Characteristics, PhD thesis, University of Florida, 2007.)

by elastic direct tunneling. Figure 6.12 shows a schematic band diagram of n-MOSFETs under mechanical stress. Applying uniaxial tensile stress shifts the ground energy level (E_0) lower in the inversion layer [34]. As a result, the tunneling probability of channel electrons at E_0 decreases because of the increased potential barrier height, while the trapping probability of tunneling electrons by oxide traps increases since their energy level shifts closer to the quasi-Fermi level (E_{FN}). Although the reduced tunneling probability (i.e., reduced I_G) decreases the noise PSD, the proximity of the quasi-Fermi level actually increases the noise PSD. However, the reduction in I_G is not dominant in determining the overall change in noise PSD. In addition, the oxide trap distribution is another important factor in determining the change in noise PSD. The induced strain in the structure inherently changes the channel carrier mobility by repopulation of carriers in the energy subbands, thereby affecting the correlated mobility fluctuations and, in the process, the noise PSD. This is further discussed later in this section.

In the conventional number fluctuation model, the drain current noise PSD can be written as [24]

$$S_{I_D}(f) = kTWL\frac{N_t(E_{Fn})\lambda\kappa^2 I_D^2}{f^\gamma N^2} \tag{6.41}$$

where WL is the gate area, kT is the thermal energy, N is the total number of channel carriers per unit area, λ is the tunneling attenuation length in the

WKB approximation, κ is the parameter combining carrier number and correlated mobility fluctuations ($\kappa = 1$ for the number fluctuation model and $\kappa > 1$ for the unified model), and $N_t(E_{FN})$ is the trap density at the quasi-Fermi level with a unit of cm^{-3} eV^{-1}. Rewriting Equation (6.41) for applied stress, we can obtain the following [33]:

$$\ln\left(1+\frac{\Delta S_{I_D}(f;\sigma)}{S_{I_D}(f;0)}\right) = \ln\left(1+\frac{2\Delta I_D(\sigma)}{I_D(0)}\right)+\ln\left(1+\frac{2\Delta\kappa(\sigma)}{\kappa(0)}\right)+\ln\left(1+\frac{2\Delta N_t(E_{Fn};\sigma)}{N_t(E_{Fn};0)}\right)$$

$$+\ln\left(1+\frac{\Delta\lambda(\sigma)}{\lambda(0)}\right)-\ln\left(1+\frac{2\Delta N(\sigma)}{N(0)}\right)-\Delta\gamma(\sigma)\ln f \tag{6.42}$$

The above expression is further simplified since the fourth and fifth terms can be neglected. The tunneling attenuation length λ is defined in Equation (6.22) as

$$\lambda = \left[\frac{4\pi}{h}\sqrt{2m^*\Phi_B}\right]^{-1}$$

Here Φ_B is a function of stress, and the strain-induced change $\Delta\Phi_B(\sigma)$ is only a few meV for our stress level of 200 MPa, compared with $\Phi_B(0) = 3.15$ eV for conduction band electrons and 4.5 eV for valence band holes [34]. Thus, $\Delta\lambda(\sigma)/\lambda(0) \ll 1$. The total number change in channel carriers due to stress, $\Delta N(\sigma)/N(0)$, can be written in terms of the drain (I_D) and gate tunneling (I_G) currents in steady-state condition,

$$\frac{\Delta N(\sigma)}{N(0)} = \frac{I_G(0)}{I_D(0)+I_G(0)} \cdot \frac{\Delta I_G(\sigma)}{I_G(0)} \tag{6.43}$$

This quantity is also very small. Then, from Equations (6.40) and (6.41) through (6.43),

$$\ln\left(1+\frac{\Delta S_{I_D}(f;\sigma)}{S_{I_D}(f;0)}\right) = \ln\left(1+\frac{2\Delta I_D(\sigma)}{I_D(0)}\right)+\ln\left(1+\frac{2\Delta\kappa(\sigma)}{\kappa(0)}\right)$$

$$+\ln\left(1+\frac{\Delta N_{t,eff}(E_{Fn};\sigma)}{N_{t,eff}(E_{Fn};0)}\right)-\Delta\gamma(\sigma)\ln f \Rightarrow \ln\left(1+\frac{\Delta S_{I_D}(1Hz;\sigma)}{S_{I_D}(1Hz;0)}\right)$$

$$= \ln\left(1+\frac{2\Delta I_D(\sigma)}{I_D(0)}\right)+\ln\left(1+\frac{2\Delta\kappa(\sigma)}{\kappa(0)}\right)+\ln\left(1+\frac{\Delta N_{t,eff}(E_{Fn};\sigma)}{N_{t,eff}(E_{Fn};0)}\right) \tag{6.44}$$

The magnitude change in noise PSD due to strain has been expressed with three terms in Equation (6.44). The trap density $N_t(E_{Fn})$ should change with strain. The effective change in trap density $(\Delta N_{t,eff}/N_{t,eff})$ is accounted for in Equation (6.44), which is due to strain-induced trapping position change in energy space, and is also related to spatial trap distribution. For extracting the stress dependence of the exponent γ, it is assumed that the oxide traps are distributed exponentially over energy (E) and space (z). Also, a continuous distribution of traps along the oxide depth direction is assumed for ultrathin gate oxide with large gate areas. The trap density is represented as [33]

$$N_t(E_0, z; \sigma) = N_{t0}(\sigma) \exp\left[\xi(\sigma).q\left(|V_{ox}|\frac{z}{t_{ox}} - \Delta\Phi_B(\sigma)\right) + \eta(\sigma)z\right] \quad (6.45)$$

where the terms in the exponential argument are due to oxide band bending, applied stress, and spatial trap distribution, respectively. In general, the parameters N_{t0}, ξ, and η are functions of mechanical stress since applied stress can alter the trap distribution by affecting both trap energy and existing interface strain between the Si channel and the oxide. The signs for ξ and η are positive (negative) for the exponential increase (decrease) for increasing distance from the interface and increasing energy above the Si band edge. For clarification, we also state the signs of $\Delta\Phi_B(\sigma)$, that is,

$$\Delta\Phi_B(\sigma) = \begin{cases} >0 \Rightarrow \text{for n channel MOSFET under tensile stress and} \\ \text{p channel MOSFET under compressive stress} \\ <0 \Rightarrow \text{for n channel MOSFET under compressive stress and} \\ \text{p channel MOSFET under tensile stress.} \end{cases} \quad (6.46)$$

These signs of $\Delta\Phi_B(\sigma)$ reflect the ground energy-level shifts in the inversion layer for applied different types of stresses. Trapping by channel carriers in higher energy levels is neglected since the contribution to noise PSD is much smaller. The integral form of the drain current $1/f$ noise PSD in the charge trapping model is written similar to Equation (6.21) as [24]

$$S_{I_D}(f) = WL\frac{I_D^2}{N^2}\kappa^2\int_{E_{Vox}}^{E_{Cox}} 4N_t(E,z)f_t(E)(1-f_t(E))dE\int_0^{t_{ox}}\frac{\tau(E,z)}{1+(2\pi f\tau(E,z))^2}dz \quad (6.47)$$

where f_t is the trap occupation function, τ is the trap time constant, and E_{Cox} and E_{Vox} are the oxide conduction and valence band edges, respectively. The expression is valid for a low drain bias [24]. Equation (6.47) can be rewritten for applied mechanical stress, using Equation (6.45),

$$S_{I_D}(f;\sigma) = WL\frac{I_D(\sigma)^2}{N^2}\kappa^2(\sigma)\exp[-\xi(\sigma)q\Phi_B(\sigma)]\int_{E_{V_{ox}}}^{E_{C_{ox}}} 4N_{t0}(\sigma)f_t(\sigma)(1-f_t(\sigma))dE$$

$$\cdot\int_0^{t_{ox}}\frac{\tau(\sigma)\exp[(\xi(\sigma)q\,|\,V_{ox}\,|\,/t_{ox}+\eta(\sigma)).z]}{1+(2\pi f\tau(\sigma))^2}dz \tag{6.48}$$

where

$$f_t(\sigma) = \left[1+\exp\left(\frac{E-E_{Fn}-q\Delta\Phi_B(\sigma)}{kT}\right)\right]^{-1}\ and,\tau(\sigma) = \tau_0\exp[z\,/\,\lambda(\sigma)]$$

The stress-dependent trap occupation function, $f_t(\sigma)$, is introduced to describe the stress dependence of the trapping probability of tunneling channel carriers by oxide traps.

One of the dominant factors affecting the noise magnitude is change of trap location in energy space. This factor influences the noise PSD depending on the relative distance from the quasi-Fermi level, and it is possible the detrimental effects of this can be avoided to some extent by proper choice of gate bias or strain engineering. More specifically, the quantisation effective mass that determines the lowest energy level in the inversion layer is approximately three times larger for electrons than for holes. At a relatively lower gate bias, compared to p-MOSFETs, n-MOSFETs can be biased such that the ground energy levels are located below the Fermi level, and thus the $1/f$ noise PSD can be reduced by applied stress. $1/f$ noise PSD magnitude is reduced for both n-MOSFETs under tensile strain and p-MOSFETs under compressive strain due to energy distribution of oxide traps. The stress-altered channel mobility, $\mu_{eff}(\sigma)$, is another key contributor to the noise PSD change, especially at low gate and drain biases. In long-channel devices, the noise PSD magnitude change can be approximately related to the drain current change as [33]

$$\Delta S_{I_D}(1Hz;\sigma)\,/\,S_{I_D}(1Hz;0) \cong 4\Delta I_D(\sigma)\,/\,I_D(0) \tag{6.49}$$

6.6 Noise in Strain-Engineered MOSFETs

MOS transistors generally show higher $1/f$ noise than bipolar transistors, and are therefore usually less preferred in low-noise applications. CMOS technology, on the other hand, is superior in terms of low cost, scalability, and low power. However, the $1/f$ noise of the MOS transistors is a problem that must be taken care of. With downscaling of device dimensions $1/f$ noise increases, which makes it extremely important not only to understand the origin of the noise, but also to reduce the noise magnitude for accurate detection of the desired signal. The $1/f$ noise is sensitive to technology; the choice of gate oxide material and oxidation/deposition process, as well as channel type and material, can have a large impact on the noise performance. The trap density and Hooge parameter can both be used as figures of merit for the $1/f$ noise performance, irrespective of the origin of the noise. In this section, low-frequency noise in advanced and highly scaled MOS transistor architectures is discussed.

6.6.1 Low-Frequency Noise Measurements

The noise measurement setup includes an Agilent E5263A two-channel high-speed source monitor unit (SMU), a SR 570 LNA, and an Agilent 35670A dynamic signal analyser, as shown in Figure 6.13. The SMU provides the necessary drain and gate bias, the minute fluctuations in the drain source voltage are amplified using the LNA, and the output of the LNA is fed to the dynamic signal analyser that performs the fast Fourier transform on the time-domain signal to yield the voltage noise power spectral density (S_V) in the 1–100 kHz range after correcting for preamplifier gain. Both the time-domain signal and its running Fourier transform can be studied using the signal analyser. DC measurements are done by using the

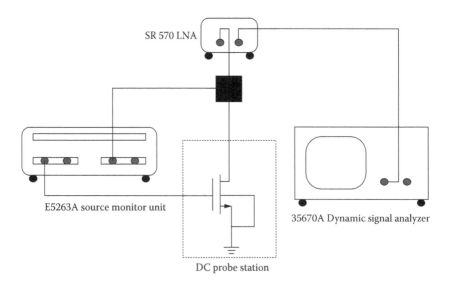

FIGURE 6.13
Experimental setup for noise measurements.

biasing network and supplying appropriate drain and gate bias using the Easy-Expert software. The minute fluctuations in the drain/source voltage were amplified to the measurable range using low-noise SR570 preamplifier operating in the 0.03 Hz–300 kHz range with voltage gain set to 1,000. In order to obtain a stable spectrum, the number of averages was set at 30 and a 90% sampling window overlap was used for optimal real-time processing. A computer interface was provided through a General Purpose Interface Bus (GPIB) connection to control the dynamic signal analyser and automate the noise data collection.

6.6.2 Strained Si MOSFETs

Figure 6.14 shows the typical drain voltage $1/f$ noise spectra as a function of gate voltage of a p-MOSFET with pseudomorphic strained Si grown on a fully relaxed SiGe buffer layer of dimension L/W = 300 µm/100 µm (strained Si on an 18% Ge buffer layer). The $1/f'$ noise spectrum is shown for four different gate voltages, which depicts increase in noise magnitude with gate voltage. The γ is found to be about 1.01, which is taken as almost $1/f$ nature for the samples at all gate voltages. Trap density is related to the input-referred flicker noise by the relation given as

$$S_{V_G}(f) = \left(\frac{q}{C_{ox}}\right)^2 \frac{1}{WLf^{\gamma}} \frac{N_T(E_F)}{\lambda}, S_{I_D}(f) = g_m^2 S_{V_G}(f) \qquad (6.50)$$

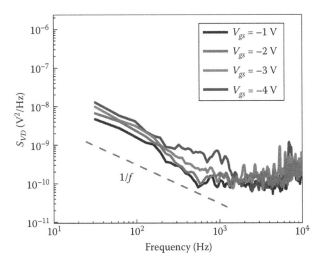

FIGURE 6.14
Drain voltage $1/f$ noise spectra as a function of gate voltage of a p-MOSFET with pseudomorphic strained Si grown on fully relaxed SiGe buffer layer of dimension L/W = 300 µm/100 µm (strained Si on an 18% Ge buffer layer).

where $N_T(E_F)kT$ is the interface state density (D_{it}) per unit energy at the Fermi energy level, and λ is McWhorter's tunneling parameter [35]. Increasing gate bias increases the number of active traps, and therefore increases the noise level. Variability of $1/f$ noise is largely due to the statistics of trap density and location. It was proposed that at low bias, nonuniform carrier densities can be formed, resulting in a change of current path and increased noise variability. The general model for describing a relation between inversion charge, N_{inv}, and the Coulomb scattering parameter for the contribution to mobility fluctuation of the charged traps, α_{sc} (with $\alpha_{sc,i}$ being the contribution of the ith trap and μ_{c0} is 5.9×10^8 cm/Vs), is given as

$$\alpha_{sc} = \frac{1}{N_T} \sum_{i=1}^{N_T} \alpha_{sc,i} = \frac{1}{\mu_{c0}\sqrt{N_{inv}}} \tag{6.51}$$

As gate voltage is increased, inversion charge, N_{inv}, grows and becomes more uniform. This reduces the impact of trap location on α_{sc}, hence reducing the uncertainty in α_{sc} [36]. This can be held responsible for the bias-dependent variation in measured $1/f$ noise. Figure 6.15 shows simulated drain voltage noise spectra fitted to experimental data of the p-MOSFET, with the inset showing the noise of the device.

Figure 6.16 shows typical time-domain RTS of the p-MOSFET at different gate biases. The RTS is closely related to the origin of flicker noise. Therefore,

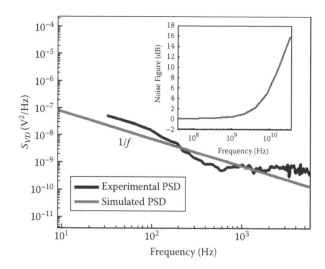

FIGURE 6.15
Comparison of simulated and experimental drain voltage $1/f$ noise spectra at a gate voltage of –5 V, for a p-MOSFET with pseudomorphic strained Si grown on a fully relaxed SiGe buffer layer of dimension L/W = 300 μm/100 μm (strained Si on an 18% Ge buffer layer).

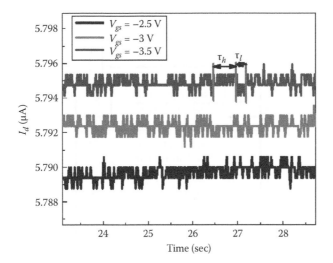

FIGURE 6.16

Typical time-domain RTS of the p-MOSFET at different gate biases, with emission and capture time constants shown in the figure.

the voltage dependencies can be similarly explained by the increase in active traps with gate voltage. The mean low time, τ_l (the mean duration in which current is low, i.e., neutral trap state), and mean high time, τ_e (the mean duration in which current is high, i.e., charged trap state), are shown in one of the RTS.

The fabrication of the p-MOSFET device [37] included a step-graded buffer layer, which is grown by gas source MBE (Daido Sanso VCES2020) at 800°C. The starting material consists of a 3 in. diameter, P-type, 5–10 Ω-cm, Si (100) wafer with 5,000 Å Si buffer, 2.1 µm step-graded SiGe buffer (0–18% Ge in seven steps), and 0.9 µm $Si_{0.82}Ge_{0.18}$ buffer cap layer. All epitaxial layers were unintentionally doped p-type to 10^{16} cm^{-1}. Si wafers with an epilayer (thickness 0.5 pm) on an n- Si (100) substrate were processed along with the strained Si wafers to act as controls. Figure 6.17 shows the schematic diagram of the p-MOSFET. The SiGe buffer and strained h layer are grown at 800 and 700°C, respectively. The strained Si epilayer (180 Å) is thermally oxidised at 700°C to form 100 Å gate oxides. The p-type doping of the SiGe buffer and strained Si layer results in a depletion mode device. At low gate bias, the confined holes at the strained Si/SiGe buffer interface dominate channel conduction and a buried channel device is formed. As the gate bias is increased, the increment in potential due to charge carriers in the parasitic channel provides a forward bias to the surface channel. Eventually at large bias a surface channel device is set up and the buried channel is suppressed because of the degenerative action of the surface channel field. Figure 6.18 shows the typical I_d-V_d characteristic of the p-MOSFET.

Figure 6.19 shows a similar drain voltage noise power spectral density of a strained Si/SiGe n-MOSFET device observed in dynamic signal analyser

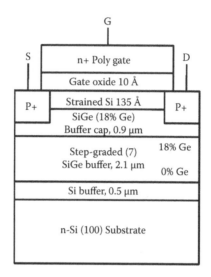

FIGURE 6.17
Schematic of a p-MOSFET on pseudomorphic strained Si grown on fully relaxed SiGe buffer layer. (After Bera, L. K., Studies on Applications of Strained Si for Heterostructure Field-Effect Transistors (HFETs), PhD thesis, Indian Institute of Technology, Kharagpur, 1998 [38].)

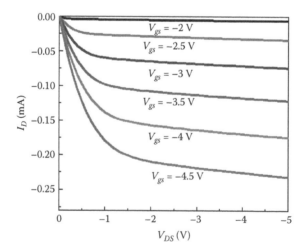

FIGURE 6.18
Typical I_d-V_d characteristics of a p-MOSFET with pseudomorphic strained Si grown on a fully relaxed SiGe buffer layer of dimension $L/W = 300\,\mu m/100\,\mu m$ (strained Si on an 18% Ge buffer layer).

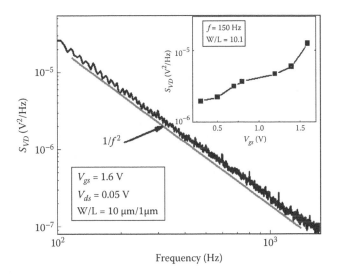

FIGURE 6.19
Drain voltage noise power spectral density of strained Si n-MOSFET with W/L = 10 µm/1 µm at V_{ds} = 0.05 V and V_{gs} = 1.6 V.

Agilent 35670A, with the inset showing the gate voltage dependency of noise at a frequency of 150 Hz. The PSD resembles the $1/f^2$ nature of the Lorentzian spectrum. The inset of Figure 6.19 shows the variation for a particular frequency of 150 Hz and compares the drain current noise power spectral densities for three n-MOSFETs with dimensions W/L = 10 µm/1 µm, W/L = 10 µm/1 µm, and W/L = 5 µm/1 µm.

The flicker noise shows a Lorentzian nature. We have also observed similar increase in noise level with gate voltage. Figures 6.20 and 6.21 show this variation of drain current noise power spectral density, observed for a fixed drain voltage of 50 mV for a device with W = 10 µm and L = 1 µm. It is observed that the noise level is higher in the device with the MOSFET with higher width, although with gate length increasing, the noise level reduces rapidly. The correlated mobility fluctuation model is given by [24] (a similar form is shown in Equation (6.37))

$$S_{I_D}(f) = \frac{kTI_D^2}{\lambda f^{\gamma} WL}\left(\frac{1}{N} \pm \lambda_{sc}\mu_{eff}\right)^2 N_t(E_{fn})$$

(6.52)

where S_{I_D} is the transistor's current noise spectral density, k is Boltzmann's constant, T is the temperature in Kelvin, λ is the tunneling parameter, f is the frequency, γ is the characteristic exponent, W and L are the active device's channel width and length, respectively, N is the total number of charge carriers in the channel, λ_{sc} is a scattering parameter, μ_{eff} is the effective electron

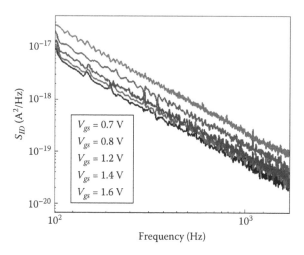

FIGURE 6.20
Variation of drain current noise power spectral density of strained Si n-MOSFET with
W/L = 10 μm/1 μm at $V_{ds} = 0.05$ V.

mobility, N_t is the volume oxide trap density, and E_{fn} is the quasi-Fermi level.
By substituting the drain current expression, Equation (6.52) can be rewritten as

$$S_{I_D}(f) = \frac{kTW}{\lambda f^\gamma L^3} (q \pm \lambda_{sc}\mu_{eff}Q_s)^2 N_t(E_{fn}) \qquad (6.53)$$

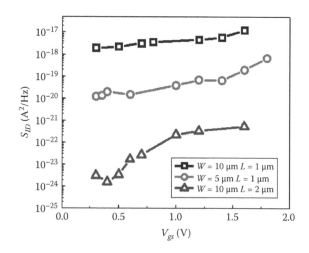

FIGURE 6.21
Variation of drain current noise power spectral density for three devices with W/L =
10 μm/1 μm, W/L = 5 μm/1 μm, and W/L = 10 μm/2 μm at $V_{ds} = 0.05$ V and $f = 150$ Hz.

where Q_s represents the area charge density of the conducting carriers in the channel. Equation (6.53) shows that the drain current noise power density is proportional to W/L^3, as observed in our results. From Equation (6.53), it can be deduced that even if one scales down W and L by the same proportion such that the W/L ratio remains constant, the drain current noise would still increase by $1/L^2$ with technology downscaling. This raises a concern about the $1/f$ noise performance of future generations of ultra-deep submicrometer MOSFETs, where low drain current noise is desired.

It is known that the $1/f$ noises of n-MOS and p-MOS transistors tend to show different gate voltage dependencies [39]. There are two existing explanations for these differences. The first one says the noises of n-MOS and p-MOS transistors are two different mechanisms: surface trapping mechanism for n-MOS transistors and bulk mobility fluctuation mechanism (Hooge's model) for p-MOS transistors. The second one says the noises of both n-MOS and p-MOS transistors are due to trapping, with the trap density constant for n-MOS transistors and varying with gate voltage for p-MOS transistors [39].

Multiple-level RTN is observed in the measured devices. In addition to this, in one set of devices, both a fast- and a slow-varying RTN (with small and large time constants, respectively) are observed, indicating the existence of both fast and slow traps, as shown in Figure 6.22. The emission and capture times of the slow RTN (approximated as a simple two-level RTN with a single active trap) are calculated as 6.920 and 3.266 ms, respectively. The variation of emission and capture times is observed by varying the gate bias from 0 to 300 mV for the fast-varying RTN. From the emission and capture

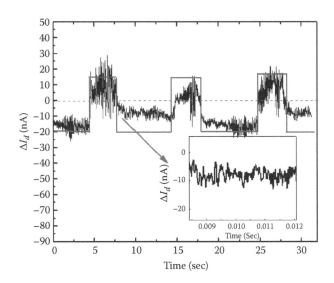

FIGURE 6.22
Slow-varying approximated RTN, with inset showing fast-varying RTN with W/L = 5 μm/5 μm and V_{gs} = 100 mV.

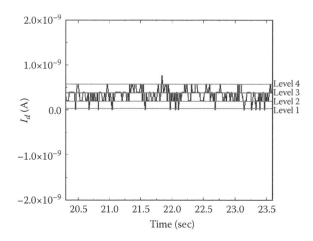

FIGURE 6.23
Four-level fast-varying random telegraph signal.

time measurements, the positions of the traps are also determined using Equation (6.36). The fast-varying RTN is modeled as a four-level RTN originating from two traps, and the trap depths are calculated as 1.374 and 14.5 Å inside the oxide. The second trap is clearly a slow-varying trap and is located deeper in the oxide layer.

In another device with W/L = 5 μm/5 μm, a four-level RTN is observed, which needs to be modeled by considering three simultaneous traps. This is shown in Figure 6.23. Since the level where the trap is full corresponds to the high current level, all the traps are acceptor type. Thus level 1 corresponds to the state where all three traps are empty. Again, one of the traps (let us call this trap 1) is responsible for *multiple-electron trapping* simultaneously. This can be assumed since there is no transition form level 1 to level 2; however, there are transitions from level 1 to level 3 and level 1 to level 4 and vice versa. Thus, level 2 corresponds to trap 1 partially full, trap 2 empty, and trap 3 empty; level 3 corresponds to trap 1 full, trap 2 full, and trap 3 empty; and level 4 corresponds to all three traps full.

The emission and capture times at V_{gs} = 100 mV as calculated from Figure 6.16 are shown in Table 6.1.

TABLE 6.1

Calculated Mean Capture and Emission Times

Trap	Mean Capture Time (τ_c)	Mean Emission Time (τ_e)
Trap 1	210 ms	16 ms
Trap 2	52 ms	19 ms
Trap 3	21 ms	109 ms

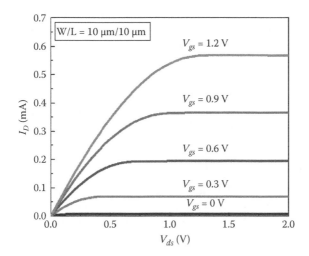

FIGURE 6.24
Typical I_d-V_{ds} characteristics of a strained Si/SiGe n-MOSFET device.

The n-MOSFETs in this study were fabricated in the Si CMOS process. The strained Si n-MOSFET substrate was 1.5 μm Si (100), 75 nm completely relaxed (19% Ge) with 800 nm graded SiGe buffer and epitaxially grown ultra-thin (150–200 Å) strained Si layer. A boron well implant of dose 2×10^{12} is used with energy 120 keV. A channel implant of dose 2×10^{12} and energy 20 keV is used. About 3.5 nm gate oxide is thermally grown on strained Si layer. Polysilicon gate is deposited of thickness ~4,000 Å. $POCl_3$ dopant diffusion is done for 60 min at 800°C. RTA is done at 950°C for 1 s, followed by surface cleaning with $HF:H_2O$ for 10 s. The shallow S/D arsenic implant is done with a dose of 2×10^{15} at 15 keV. Body contact implant of BF_2 is done at 25 keV with a dose of 3×10^{15}. Finally, the gate metal contact Ti/TiN/AlSiCu of thickness 8,500 Å is deposited, thus forming a Si/strained Si/relaxed SiGe MOS structure. Figure 6.24 shows the I_d-V_{ds} characteristics of strained Si channel n-MOSFETs with gate length of 10 μm and gate width of 10 μm.

6.7 Noise in Multigate FETs

Due to its better gate control, and substrate bias-independent threshold voltage, the multigate (MG) and gate-all-around (GAA) silicon nanowire transistors (SNWTs) [40] are being considered important candidates for future CMOS scaling beyond the 32 nm node. The noise margin and the inverter threshold voltage depend on the transitions between the subthreshold and strong inversion regions. High noise levels in the subthreshold region (close to the threshold voltage) may disturb the normal switching behaviours and circuit performance. SNWTs have also been widely studied as chemical and biochemical sensors [41, 42]. Biosensing by SNWTs is based on the pronounced conductance changes induced by the depletion of charge carriers in the silicon body when the charged biomolecules are bound to its surface. The high noise level in the depletion (subthreshold) region may lead to reduced signal-to-noise ratios in these sensors. This section will present low-frequency noise studied in MG and GAA devices.

6.7.1 Noise in Tri-Gate FinFET

The p-type tri-gate FinFET device was fabricated on a 1,000 Å SOI layer. The FinFET device has a fin height of 30 nm. The fin width and fin length are varied from 60 to 90 nm and 60 nm to 100 µm, respectively. A 50 Å SiO_2 gate oxide is grown. A polysilicon gate was deposited of thickness 1,500 Å, followed by pocket implantation and a 300 Å SiO_2 spacer deposition. A deep source/drain implant is followed. Silicidation is done for contact formation by 300 Å Ni/400 Å TiN depositions with RTA at 550–600°C for 1 min. Seven hundred angstroms of 1 µm thick Al pad is deposited for contact formation. The final device is annealed in forming gas at 420°C for 30 min. Figure 6.25 shows the typical I_d-V_{ds} characteristics of the p-type tri-gate FinFET. Typical device dimensions used for measurements are fin length of 160 nm, width of 60 nm, and oxide thickness of ~5 nm. The device displays excellent performance in terms of near-ideal subthreshold slope (SS) (~64–72 mV/dec) and high I_{on}/I_{off} ratios (~10^6). The device structure simulated in SILVACO's

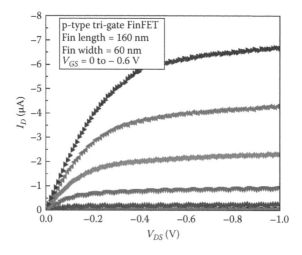

FIGURE 6.25

Typical I_d-V_{ds} characteristics of the p-type tri-gate FinFET with fin length 160 nm, gate width of 60 nm, and gate height of 30 nm.

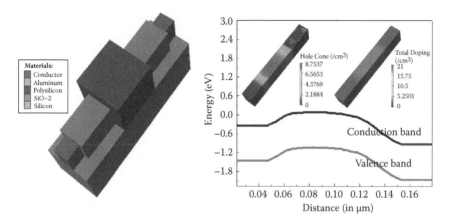

FIGURE 6.26
(a) Simulated device structure of the tri-gate FinFET. (b) The energy and diagram of the tri-gate FinFET with the inset showing the hole concentration and total doping along the channel.

ATLAS framework is shown in Figure 6.26(a). The band diagram is shown in Figure 6.26(b) with the insets showing hole concentraion and total doping along the channel.

Figure 6.27 shows the measured $1/f^t$ drain voltage noise spectra of the tri-gate device as a function of frequency at different gate voltages. Figure 6.28 shows the fit of simulated drain voltage noise spectral density, obtained from TCAD device simulation in SILVACO, with the experimental data at a gate

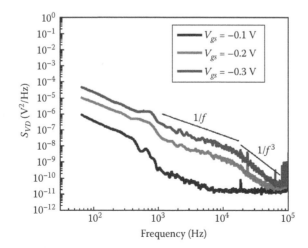

FIGURE 6.27
Measured $1/f^t$ drain voltage noise spectra of the tri-gate device as a function of frequency at different gate voltages and a drain bias of –0.05 V.

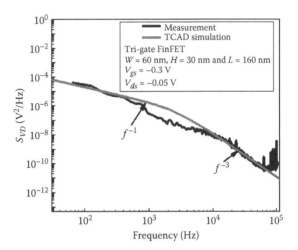

FIGURE 6.28
Simulated drain voltage $1/f^t$ noise spectra fitted with the measured spectra at a gate bias of –0.3 V, and drain voltage of –0.05 V for the tri-gate FinFET device.

voltage of –3 V, which reveals the noise to be $1/f$ in nature and then becoming $1/f^3$ type.

The gate voltage dependency of the drain voltage noise power spectral density is shown in the Figure 6.29 as a function of gate voltage at two different frequencies, which shows a constant increase in the noise level with

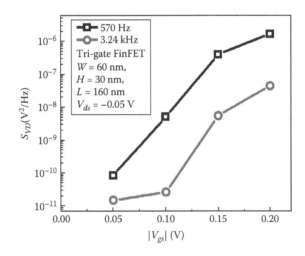

FIGURE 6.29
The gate voltage dependency of the drain voltage noise power spectral density at two different frequencies.

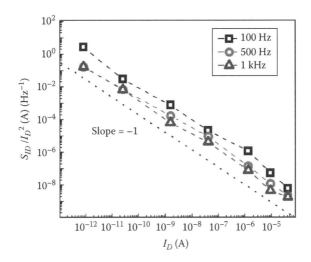

FIGURE 6.30
Drain current noise power spectral density, S_{ID}/I_D^2, as a function of drain current, I_D, at three different frequencies, which shows a slope of –1.

gate bias. This is also due to the the increase in trap densitiy with gate bias as depicted by Equations (6.51) and (6.52). Figure 6.30 shows the drain current noise power spectral density S_{ID}/I_D^2 as a function of drain current I_D at three different frequencies, which shows a slope of –1. This can be explained from Equation (6.30) as S_{ID}/I_D^2 being proportional to $1/I_D$ for drain current, and the total charge density Q_i being independent of $V_{DS} \gg mkT/q$. The noise observed is thus mobility $1/f$ noise.

The drain current RTSs of the FinFETs observed are shown in Figure 6.31 in a time frame of 10 s at five different gate voltages, depicting an increase in the mean low time of the current level and decrease in the mean high time. The capture and emission times are identified on the time-domain signal, which leads to an interesting observation. It is known that a charged trap leads to increased resistance, thereby leading to a high level in current. Also, with increased gate bias, the trap occupancy should increase, or the emission time should increase. Depending on these, the high current level is identified as capture time and the low current level as the emission time. This leads to the identification of the active trap to be a donor trap (in the charged state the trap is empty).

The capture and emission times of the tri-gate FinFET obtained from Figure 6.31 are plotted against gate bias in Figure 6.32, along with the ratio of the capture and emission times. The inset shows the $\ln(\tau_c/\tau_e)$ plot as a function of gate bias with a linear fit. The results from this plot are put in Equation (6.36), and the trap depth is estimated as 0.207 nm. Figure 6.33

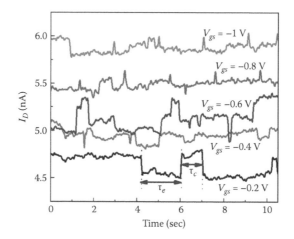

FIGURE 6.31
The time-domain RTS for a time window of 10 s, as observed in the tri-gate FinFET at different gate voltages.

shows a typical multilevel RTS observed in the FinFET showing four distinct levels. Further analysis can show that, assuming the empty state to be the low current level, we can predict that two active traps are present, although both traps are responsible for multiple-electron trapping. All traps are empty at level 1. Trap 1 is partially filled at level 2 and trap 2 is empty. At level 3 trap 1 is filled and trap 2 is partially filled. At level 4, all traps are filled.

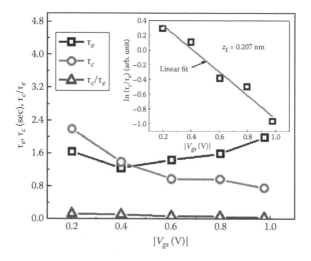

FIGURE 6.32
The capture and emission times of the tri-gate FinFET, calculated from the RTS; inset shows the $\ln(\tau_c/\tau_e)$ plot as a function of gate bias with a linear fit. The trap depth is estimated as 0.207 nm.

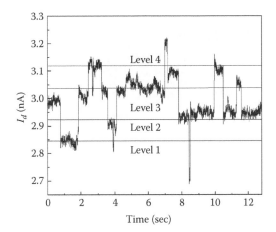

FIGURE 6.33
A typical multilevel RTS observed in the tri-gate FinFET.

6.8 Noise in Silicon Nanowire Transistors (SNWTs)

Two approaches are generally used to fabricate Si NWs as well as other semiconductor NWs: bottom-up and top-down. In the first method, NWs are usually grown using a metallic catalyst on a separate substrate, usually through a vapour-liquid-solid (VLS) growth mechanism. After a chemical or mechanical separation step, the NWs are harvested and transferred to another substrate. In the top-down approach, the NWs are fabricated using a CMOS-compatible technology, such as lithography-based patterning and etching. Unlike the bottom-up approach, where the NWs are randomly distributed, the top-down method enables accurate positioning of the NWs across the wafer and facilitate the ultra-large-scale integration for high-performance nanoelectronic circuits. Moreover, due to process difficulties related to the length of grown NWs, NW release, and gate etch process, most of the VLS-grown NW transistors have omega-shaped gate (Ω-gate) geometry and are thus not full gate-all-around. In a long-channel MOSFET, carriers encounter various scattering mechanisms on their path toward the drain terminal. Carrier mobility is a well-known benchmark to judge the intrinsic performance of a long-channel MOSFET. Equation (6.1) indicates that the injection velocity near the source determines the on-state current of a short-channel device. State-of-the-art short-channel devices do not operate in the fully ballistic regime (they are at roughly 60% of the ballistic limit), and mobility is related to velocity through effective mass and ballistic ratio. Therefore, understanding the carrier mobility is beneficial to design and engineer new devices for future CMOS generations. The presence of significant resistive and capacitive parasitics, as well as a lack of large capacitance

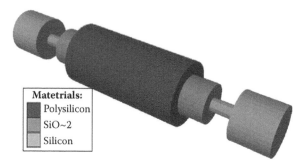

FIGURE 6.34
Simulated schematic structure of the Si-nanowire.

due to the small size of the NW channel, adds complexities to the extraction of the intrinsic NW characteristics. The gate-all-around NW structure, simulated in SILVACO's 3D device simulation module, is shown in Figure 6.34.

Figure 6.35 shows the frequency dependence of the measured drain current noise power spectral density S_{Id} of six samples of 90 nm p-type SNWTs, biased at $V_{ds} = -50$ mV and $I_d = 3.1$ nA. A typical $1/f^\gamma$ behaviour with $\gamma = 1.03$ is shown. The noise magnitude dispersion shown in the inset of Figure 6.35 can be due to the lattice quality and mobility variations of the ultra-scaled dimension of SNWTs. Noise measured on 20 different samples with identical gate length is averaged.

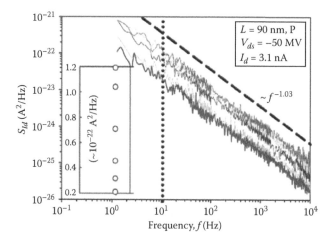

FIGURE 6.35
Drain current noise spectral density S_{Id} of six individual p-type SNWTs with L = 90 nm biased at $V_{ds} = -50$ mV at constant $I_d = 3.1$ nA. Measured S_{Id} dispersion at $f = 10$ Hz is shown in the inset. (After Wei, C., Y.-Z. Xiong, X. Zhou, N. Singh, S. C. Rustagi, G. Q. Lo, and D.-L. Kwong, *IEEE Electron Dev. Lett.*, 30, 668–671, 2009. With permission [43].)

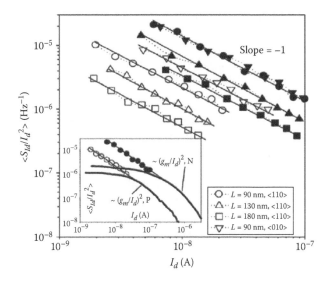

FIGURE 6.36
Average normalised drain current spectral density at $|V_{ds}| = 50$ mV and $f = 10$ Hz vs. drain current for both n- and p-type SNWTs with $L = 90$, 130, and 180 nm, respectively. The noise data of 90 nm n- and p-type SNWTs are compared with the corresponding (constant $\times$ $(g_m/I_d)^2$) in the inset. Open symbols, p-type; solid symbols, n-type; solid line, fitted by Equation (6.63). (After Wei, C., Y.-Z. Xiong, X. Zhou, N. Singh, S. C. Rustagi, G. Q. Lo, and D.-L. Kwong, *IEEE Electron Dev. Lett.*, 30, 668–671, 2009. With permission [43].)

Figure 6.36 shows the average normalised drain current noise spectral density $\langle S_{Id} / I_d^2 \rangle$ at $f = 10$ Hz as a function of the drain current I_d at constant V_{ds} of −50 mV, with varying gate voltage for both n- and p-type SNWTs with different gate lengths and channel orientations. The normalised variations S_{Id}/I_d^2 vs. I_d exhibit a slope close to −1, which shows that the mobility fluctuation $1/f$ noise model is dominant according to Equation (6.29). If the $1/f$ noise is due to the carrier number fluctuations, S_{Id}/I_d^2 would have been independent of the drain current, and S_{Id}/I_d^2 becomes proportional to $(g_m/I_d)^2$ per Equation (6.20). In the inset of Figure 6.36, a large slope deviation between S_{Id}/I_d^2 and $(g_m/I_d)^2$ in the subthreshold region further confirms the mobility fluctuation mechanism. A similar form of Equation (6.26) depicting the empirical relation of the mobility fluctuation model [13] is given by

$$\frac{S_{Id}(f)}{I_d^2} = \frac{\alpha_H}{fN} \qquad (6.54)$$

with α_H being Hooge's parameter and N the total number of carriers under the gate. In the subthreshold region N is very small, and in SNWT it is typically less than 20. The study is centred on the $1/f$ noise in the subthreshold region.

The subthreshold current of lightly doped GAA SNWTs can be expressed as [44]

$$I_d = \mu \frac{\pi R^2}{L} n_i kTe^{\frac{q(V_{gs} - \Delta\Phi)}{kT}} \left(1 - e^{-\frac{qV_{ds}}{kT}} \right) \tag{6.55}$$

where μ is the effective mobility, R is the radius of the gate-all-around SNWT, L is the channel length, n_i is the intrinsic carrier concentration, and $\Delta\Phi$ is the work function difference between the gate electrode and the almost intrinsic silicon body.

From the expression of surface potential, ψ_s at the Si/SiO$_2$ interface, a β-dependent expression $f(\beta)$ is defined by Jimenez et al. as [44]

$$f(\beta) = \frac{q}{kT} \left(V_{gs} - V_o - V(y) \right) \tag{6.56}$$

with

$$V_o = \Delta\Phi + \frac{kT}{q} \ln \left(\frac{8kT\varepsilon_{Si}}{q^2 n_i R^2} \right) \tag{6.57}$$

where $V(y)$ is the electron quasi-Fermi potential at y along the channel and β is a parameter related to R and ψ_s. The carrier charge density per unit area of the channel, $Q_i(y)$ at location y, is given in the subthreshold region as [44]

$$Q_i(y) \approx 4 \frac{\varepsilon_{Si}}{R} \frac{kT}{q} (1 - \beta) = \frac{q n_i R}{2} e^{\frac{q}{kT}(V_{gs} - \Delta\Phi - V(y))} \tag{6.58}$$

Following the approach described by [28] and using Equations (6.55) and (6.58), the final expression of S_{Id}/I_d^2 in the subthreshold region from the mobility fluctuation model for SNWTs is given as

$$\frac{S_{Id}(f)}{I_d^2} = \frac{\alpha_H}{fN} = \frac{\alpha_H}{f} \frac{2\mu kT}{L^2} \frac{1 - e^{\frac{-qV_{ds}}{kT}}}{1 + e^{\frac{-qV_{ds}}{kT}}} \frac{1}{I_d} \tag{6.59}$$

Since the subthreshold slope of p-type SNWTs is not as ideal as n-type ones, two ideal factors m and m' related to the gate and drain biases, respectively, are added into Equation (6.55) as

$$I_d = \mu \frac{\pi R^2}{L} n_i kTe^{\frac{q(|V_{gs}| - |\Delta\Phi|)}{mkT}} \left(1 - e^{-\frac{q|V_{ds}|}{m'kT}} \right) \tag{6.60}$$

which leads to the final mobility fluctuation model in the subthreshold region as

$$\frac{S_{Id}(f)}{I_d^2} = \frac{\alpha_H}{fN} = m' \frac{\alpha_H}{f} \frac{2\mu kT}{L^2} \frac{1-e^{\frac{-q|V_{ds}|}{m'kT}}}{1+e^{\frac{-q|V_{ds}|}{m'kT}}} \frac{1}{I_d} \tag{6.61}$$

where $m = 1$ and 1.1 and $m' = 1$ and 1.03 for n- and p-type SNWTs, respectively [43]. For devices working in the ohmic region with very low drain biases $|V_{ds}|$, Equation (6.61) can be simplified as

$$\frac{S_{Id}(f)}{I_d^2} = \frac{\alpha_H}{fN} = \frac{\alpha_H}{f} \frac{q\mu}{L^2} |V_{ds}| \frac{1}{I_d} \tag{6.62}$$

The effective mobility μ has been extracted from the I-V data of the long-channel SNWTs and is around 150 and 45 cm^2/V·s for n- and p-type SNWTs, respectively, and from variations in Figure 6.36, α_H has been extracted to be $\alpha_H \approx 1.2 \times 10^{-4}$ and 7×10^{-5} for the n- and p-type SNWTs, respectively [43].

The values of the Hooge parameters, extracted by [43], are in good agreement with range for conventional silicon CMOS bulk devices (SiO$_2$/polysilicon gate stack) and are also close to the values predicted from the ITRS road map for the 45 nm technology node [45]. From Equation (6.62), it is clear that S_{Id} is proportional to μ in the channel for a given V_{ds} and I_d. Also, at a fixed-bias condition, S_{Id} is proportional to μ^2 (from Equations (6.61) and (6.62)).

6.9 Noise in Heterojunction Bipolar Transistors

With rapid device downscaling, low-frequency noise in transistors is becoming a very dominating criterion for device design. Especially in RF applications, the presence of low-frequency noise as undesirable phase noise is very critical for circuit designing. Heterojunction bipolar transistors (HBTs) in SiGe:C technology are becoming an important candidate with quite a remarkable combination of RF performance and ruggedness. These HBT devices have higher cutoff frequency (f_T) and current gain (β) over their identical Si counterparts, and incorporation of carbon induces base width reduction. However, the complex fabrication process and induced strain largely affect the low-frequency noise components of these devices to a great extent. Thus, the characterisation of low-frequency noise in HBT devices has become of immense importance.

Low-frequency (LF) noise consists of mainly two components, flicker noise (1/f noise) and random telegraph noise (RTN). Various previous researchers have explored the LF noise in bipolar junction transistors (BJTs), especially the source of $1/f$ noise in BJTs [46, 47]. LF noise characteristics of SiGe HBTs are also reported [48]. One of the most important sources of $1/f$ noise is located within the thin SiO_2 interfacial layer between the monosilicon and polysilicon emitter regions. The physical mechanism of this noise is interpreted by tunneling probability fluctuation [47] or number carrier fluctuation [46], and it is still unclear.

The random telegraph noise (RTN) observations, on the other hand, in BJTs, have mainly been interpreted in terms of noise sources in the base-emitter space charge region [49] and interfacial oxide in the emitter region [50]. In the former case, the traps in the base-emitter space charge region cause the barrier height across the junction to fluctuate, resulting in RTS pulses. In the latter case, the increase of nonideal base current gives rise to RTS pulses, with the source being the traps in the spacer oxide at the emitter interface. Several physical mechanisms exist, such as fluctuating barrier height and trap-assisted tunneling current across the junction, multiphonon capture by traps, fluctuating recombination rate in the base-emitter space charge region, and fluctuating capture cross sections. Overall, the trapping/de-trapping mechanism in bipolar junction transistors is yet to be fully understood. Studies on low-frequency noise in SiGe:C HBT are scarce and exact physical mechanisms are vague. In this work, low-frequency noise behaviour consisting of $1/f$ noise and RTN in SiGe:C high-speed HBT has been characterised. Bias dependency of $1/f$ noise in the base contact is demonstrated, and existence of a nonideal base current component has been shown. RTN is characterised in terms of amplitude and characteristic times in the low and high level of base current as a function of bias. Mechanisms behind the bias dependency of RTN are explained based on the experimental results with the help of fluctuating barrier height theory.

The SiGe:C HBT was fabricated in an industry standard 0.25 µm BiCMOS process with 25 lithographic steps. The HBT devices used in the work have an emitter area AE of 0.42×0.84 µm^2 with the typical parameters $f_T/f_{max} = 75/90$ GHz, $BV_{CEO} = 2.4$ V, and $\beta = 100$. The HBT SiGe base layer incorporates a very low carbon concentration of about 1×20 Cm^{-3} to suppress boron diffusion in the base layer and the surrounding Si regions. Further details on the device can be found in [51].

Figure 6.37(a) shows the typical output characteristics (I_C-V_{CE}) of the NPN SiGe:C HBT with an emitter area A_E of 0.42×0.84 µm^2. The typical gummel plot of the HBT is shown in Figure 6.37(b), with the inset showing the current gain β vs. V_{BE} plot. Apart from this, an f_{max} of 90 GHz and a maximum β of 115 were measured.

6.9.1 Low-Frequency Noise Measurement of SiGe:C HBT

Figure 6.38 shows the base current noise power spectral density S_{IB} at three different base voltages, which show $1/f^2$ dependence with frequency. The

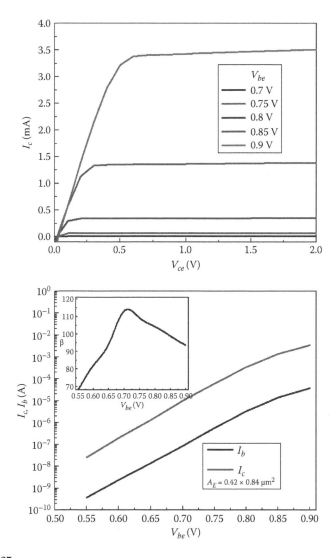

FIGURE 6.37
(a) I_C-V_{CE} characteristics of SiGe:C HBT. (b) Gummel plot of the SiGe:C HBT with emitter area $0.42 \times 0.84~\mu m^2$; inset showing the β-V_{BE} characteristics of the HBT.

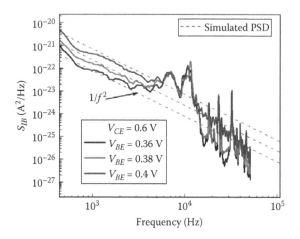

FIGURE 6.38
Base current noise power spectral density S_{IB} of the SiGe:C HBT with emitter area 0.42×0.84 μm^2 at three different values of V_{BE}.

current noise spectral density can be expressed in the following form [8], where N is the number of independent traps:

$$S_I(f) = \sum_i^N \frac{4(\Delta I_i)^2}{(\tau_{li} + \tau_{hi})((1/\tau_{li} + 1/\tau_{hi})^2 + (2\pi f)^2)} \tag{6.63}$$

where ΔI_i is the amplitude of the ith trap, and τ_{li} and τ_{hi} are the mean times in low and high current levels, respectively. Equation (6.63) has been used to calculate theoretical noise spectral densities for three different values of base bias, and they are shown in Figure 6.38 to compare with the experimental spectral density. The theoretically calculated data show good agreement with the experimental PSD with Lorentzian nature, which signifies the dominance of RTS noise in the low-frequency noise in the low-bias regime.

Figure 6.39 shows the PSD at higher base currents, which reveals the $1/f$ nature of the flicker noise (FN) combined with the generation–recombination (g-r) noise and the shot noise (SN). This type of flicker noise characteristics have been reported by previous researchers, and can be expressed by the following equation:

$$S_I(f) = K_F \frac{I_B^2}{f} + \sum_i^N \frac{A_i \tau_i}{1 + (2\pi f \tau_i)^2} + 2q I_B \tag{6.64}$$

where K_F is the magnitude of the flicker noise component of the total noise measured, and A_i and τ_i are the amplitude and composite time constant of the ith G-R peak. The last term in Equation (6.64) is the shot noise component. Figure 6.40 shows the variation of S_{IB} with base current I_B for three different frequencies, which shows an I_B^2 dependence on the base current. This quadratic dependence is also often reported in SiGe HBTs [52], as

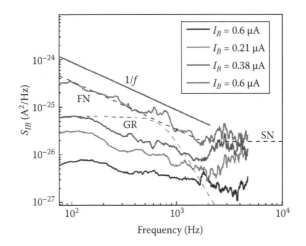

FIGURE 6.39
Base current noise power spectral density S_{IB} of the SiGe:C HBT with emitter area 0.42×0.84 μm^2 showing the low-frequency noise components, such as $1/f$, generation–recombination noise, and shot noise components at higher values of I_B.

well as in polyemitter BJTs. The $1/f$ noise amplitude K_F is given in the SPICE model as

$$S_{IB} = K_F \frac{I_B^2}{f}, \; K_B = K_F \times A_E (\mu m^2) \qquad (6.65)$$

where K_B is the figure of merit calculated as $1 \times 10^{-6} \mu m^2$.

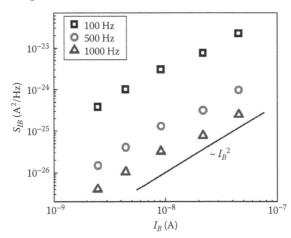

FIGURE 6.40
I_B^2 dependence of the base current noise power spectral density S_{IB} as a function of I_B for the SiGe:C HBT with emitter area 0.42×0.84 μm^2 at three different frequencies.

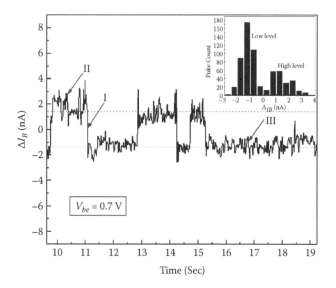

FIGURE 6.41

RTS amplitude of the SiGe:C HBT at $V_{BE} = 0.7$ V, showing three different RTN pulses: I (slow), II (fast), and III (fast). Inset shows the histogram of the pulse count vs. the amplitude.

Figure 6.41 shows a typical random telegraph signal observed at $V_{BE} = 0.7$ V depicting three different RTN pulses in the time window. These pulses are either fast or slow depending on the characteristic time constants. Pulse I is the slow RTN with a characteristic time in the order of hundreds of milliseconds. Both pulse II and III are fast-varying RTNs with the characteristic time in the order of microseconds, although pulse II and III have different amplitudes. The two distinct levels in pulse I are shown in two lines, with the inset depicting the typical histogram of a two-level random telegraph signal. The histogram plots pulse counts vs. the pulse amplitude ΔI_B, and the distinct high and low levels are shown in the figure. Figure 6.42 shows the different RTNs at different base biases, ranging from 0.6 to 0.7 V. It is noticeable how the RTN has become multilevel RTS from the regular two-level RTS, with the base bias increasing, due to the excitation of more traps. RTS amplitude scales in two different mechanisms with the base bias. In one case, the scaling is observed proportional to the total base current I_B, whereas in the other case, amplitude scales with the nonideal base current.

Figure 6.43 shows the RTS amplitude ΔI_B of a SiGe:C HBT with emitter area 0.42×0.84 μm^2 as a function of the base current I_B, which shows a linear dependence. RTS amplitude is about 1.31% of the total base current as found from the plot. The linear dependence of ΔI_B on total base current indicates the noise mechanism that affects both nonideal and ideal base currents, and the dominant noise source is located in the base-emitter space charge region. This linear dependence has been explained by a model based

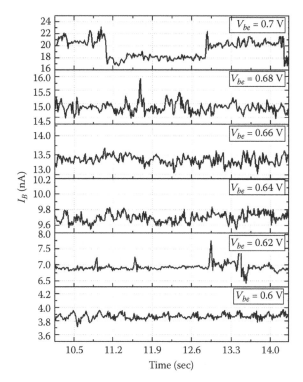

FIGURE 6.42
RTS of the SiGe:C HBT at different values of V_{BE}.

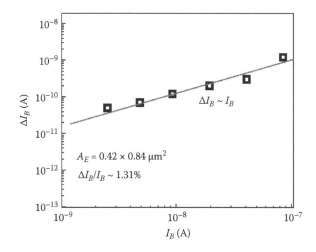

FIGURE 6.43
RTS amplitude scaling of SiGe:C HBT with the total base current I_B.

on local voltage barrier fluctuations in the emitter-base space charge region due to trapping/de-trapping of carriers as suggested by [48]. A trapped electron in the base side causes voltage barrier height across the space charge region ΔV_{BE} to reduce, which increases the base current level. RTS amplitude is given by [49]

$$\Delta I_B = \frac{L_S^2}{A_E} I_{B0} \exp\left(\frac{q\Delta V_{BE}}{kT}\right) \tag{6.66}$$

where L_S is the screening length, I_{B0} is the initial base current level, and A_E is the emitter area.

The traps in the thin oxide at the emitter interface are dominant noise sources for the low-frequency noise. The RTS amplitude due to these traps scales with the injected hole current ($\sim\exp(qV_{BE}/kT)$) and is relatively small. On the other hand, the RTS amplitude that scales with the nonlinear base current shows a weaker bias dependence. Figure 6.44 shows the plot of ΔI_B vs. V_{BE}, which shows a bias dependence of the nature $\sim\exp(qV_{BE}/2\,kT)$. These RTS pulses have presumably originated from the noise sources in the spacer oxide at the emitter region.

The capture and emission process of carriers in the space charge region is very complex and involves both tunneling and thermal capture, and depends on several parameters, such as temperature, electric field strength, trap energy level, and phonon energy. There are several different capture mechanisms, such as the cascade process and the multiphonon mechanism. The mean times

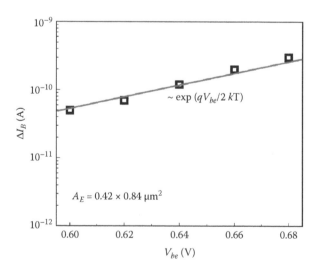

FIGURE 6.44
RTS amplitude scaling of SiGe:C HBT with the nonlinear base current component.

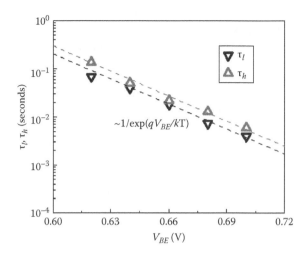

FIGURE 6.45
Characteristic times in the high and low levels of the RTS observed as a function of the bias voltage V_{BE}.

for capture and emission of an electron, τ_e and τ_c, respectively, are given by Shockley–Read–Hall statistics, as described by Equation (6.37):

$$\tau_c = \frac{1}{n v_{TH} \sigma_n}, \tau_e = \frac{\exp\left[\frac{(E_T - E_F)}{kT}\right]}{N_C v_{TH} \sigma_n}$$

where n is electron density in the vicinity of traps and depends on the location of trap in the space charge region and the bias voltage V_{BE}, v_{TH} is the thermal velocity of electrons, σ_n is the electron capture cross section of the traps, N_C is the density of state in the conduction band, and E_T and E_F are the trap energy level and Fermi potential, respectively.

The characteristic times for RTN pulse III in Figure 6.41 are shown in Figure 6.45 as a function of the base-emitter voltage. It is observed that both τ_l and τ_h decrease rapidly as $1/\exp(qV_{BE}/kT)$. A trap located near the base can capture electrons from the conduction band and holes from the valence band. Capture of carriers depends inversely on the density of the carriers in the vicinity of the trap, which can increase with bias, according to Equation (6.37), thereby reducing the mean capture time, as seen from Figure 6.45. As σ_n is electric field dependent, it can influence both the capture and emission times. The electric field decreases slightly with increasing bias voltage in the space charge region. However, this does not affect the capture cross section, and hence the characteristic times to a significant extent. Therefore, the trapping/de-trapping processes have been explained with a two-capture process (electron and hole capture), which is the most likely phenomenon for explaining the strong reduction in characteristic times with V_{BE}, as observed in Figure 6.45. It may be possible that a more complex capture process is involved. With the base region containing carbon as recombination centres, the base doping is usually very high in SiGe:C HBTs, allowing a very thin width of the base-emitter space charge layer. This thin space charge region assists tunneling of electrons across it from the neutral regions into the traps in the spacer oxide. The tunneling thickness reduces with the potential barrier as V_{BE} increases, which reflects the rapid reduction in characteristic times with the base bias. An expression for the tunneling transition to and from the traps in the space charge region or in the spacer oxide can be written as [48]

$$\tau_{tunneling} = \frac{\exp\left[2\sqrt{\frac{2meEy}{\hbar}}\right]}{4nv_{TH}\sigma_n} \tag{6.67}$$

where E is the height of the potential barrier, y is the tunneling distance, m_e is the electron effective mass, and n is the electron density in the region of tunneling. It is evident from Equation (6.67) that the characteristic times involve both two-capture and tunneling mechanisms. With emitter doping being higher than the base doping, the space charge region width will change almost entirely on the base side. This can explain the base bias dependence of at least one of the characteristic times, most likely the tunneling of holes from the base into the trap. From Equation (6.66), I_B is at the high level when an electron is trapped. So, τ_h corresponds to the mean time for a hole to be captured by the trap (the duration for which the electron remains captured in the trap), and likewise τ_l is the mean electron capture time.

6.10 Summary

The importance of low-frequency noise study and understanding of different physical mechanisms involved is emphasised. The detrimental effects of noise in RF circuits and advanced strain-engineered devices with highly scaled device dimensions are discussed in detail. Sources of low-frequency noise in the strained devices are described. Effects of strain on $1/f$ noise in strained MOSFETs are illustrated showing the diverse effects of tensile and compressive stress on the noise power spectral density. Detailed mechanisms of strain caused by applied mechanical stress on noise PSD are also discussed. Key properties include alteration of both the magnitude and exponent in the $1/f^f$ noise spectrum, resulting in larger changes in noise PSD at lower frequencies. $1/f$ noise and RTS noise study and their bias dependences are illustrated for strained MOSFETs as well as multigate FETs and nanowires. The mean capture and emission times are extracted from RTS, and their bias dependency has been demonstrated. The study of bias dependency has been demonstrated for extraction of trap locations responsible for the noise fluctuations. Low-frequency noise study in SiGe:C heterojunction bipolar transistors is also included, for which the mechanism is little different from the MOSFETs. Flicker noise characteristics and their corresponding power spectral densities have been observed, and their bias dependencies are demonstrated. The origins of random telegraph noise are depicted by recombination centres and electrically active defect states formed by carbon atoms via a generation–recombination mechanism in the base of the SiGe:C HBT. The amplitude scaling of the RTS is observed either with the total base current or with a nonideal base current component. The two-capture process and tunneling of a carrier from the neutral region are described as the fundamental noise-generating mechanisms in the bipolar transistors.

Review Questions

1. What are the fundamental noise sources in a semiconductor?
2. $1/f$ noise affects the performance of RF circuits. (True/False)
3. Noise measurements may be used as a diagnostic tool for defect determination. (True/False)
4. Addition of carbon affects the noise performance of SiGe:C HBTs. (True/False)
5. Electrically active defect states formed by carbon atoms via a generation–recombination mechanism are responsible for $1/f$ noise in SiGe:C HBTs. (True/False)
6. Thermal noise originates from the random thermal movement of electrons. (True/False)
7. The origin of $1/f$ noise in MOS transistors is due to carrier number fluctuation noise due to traps in the gate oxide. (True/False)
8. Shot noise is generated when the electrons cross the barrier independently and randomly. (True/False)
9. Generation–recombination noise in semiconductors originates from random capture or emission of carriers by localised charge centres. (True/False)
10. A mobility fluctuation noise model is used to explain the $1/f$ noise. (True/False)

References

1. International Technology Roadmap for Semiconductors (ITRS), http://public. itrs.net/.
2. T. Sakurai, Perspectives of Power-Aware Electronics (Plenary), *Proc. ISSCC Tech. Dig.*, 26–29, 2003.
3. B. Razavi, A Study of Phase Noise in CMOS Oscillators, *IEEE J. Solid-State Circ.*, 31, 331–343, 1996.
4. A. Hajimiri and T. H. Lee, A General Theory of Phase Noise in Electrical Oscillators, *IEEE J. Solid-State Circ.*, 33, 179–194, 1998.
5. M. V. Haartman, Low-Frequency Noise Characterization, Evaluation and Modelling of Advanced Si- and SiGe-Based CMOS Transistors, PhD thesis, Royal Institute of Technology (KTH), Sweden, 2006.
6. M. J. Deen and O. Marinov, Noise in Advanced Electronic Devices and Circuits, *Proc. Int. Conf. Noise Fluctuations (ICNF)*, 3–12, 2005.
7. L. K. J. Vandamme, Noise as a Diagnostic Tool for Quality and Reliability of Electronic Devices, *IEEE Trans. Electron Dev.*, 41, 2176–2187, 1994.
8. M. J. Kirton and M. J. Uren, Noise in Solid-State Microstructures: A New Perspective on Individual Defects, Interface States and Low-Frequency (1/f) Noise, *Adv. Phys.*, 38, 367–468, 1989.
9. L. K. J. Vandamme, X. Li, and D. Rigaud, 1/f Noise in MOS Devices, Mobility or Number Fluctuations? *IEEE Trans. Electron Dev.*, 41, 1936–1945, 1994.
10. W. Schottky, Über spontane Stromschwankungen in verschiedenen Elektrizitätsleitern, *Annalen Physik*, 57, 541–567, 1918.
11. A. van der Ziel, *Noise in Solid State Devices and Circuits*, John Wiley & Sons, New York, USA, 1986.
12. S. Machlup, Noise in Semiconductors: Spectrum of a Two-Parameter Random Signal, *J. Appl. Phys.*, 25, 341–343, 1954.
13. F. N. Hooge, T. G. M. Kleinpenning, and L. K. J. Vandamme, Experimental Studies on 1/f Noise, *Rep. Prog. Phys.*, 44, 479–531, 1981.
14. M. Surdin, Fluctuations in the Thermionic Current and the "Flicker Effect," *J. Phys. Radium*, 10, 188–189, 1939.
15. F. N. Hooge, 1/f Noise Sources, *IEEE Trans. Electron Dev.*, 41, 1926–1935, 1994.
16. F. N. Hooge, 1/f Noise Is No Surface Effect, *Phys. Lett. A*, 29, 139–140, 1969.
17. P. H. Handel, Fundamental Quantum 1/f Noise in Semiconductor Devices, *IEEE Trans. Electron Dev.*, 41, 2023–2033, 1994.
18. C. M. Van Vliet, A Survey of Results and Future Prospects on Quantum 1/f Noise and 1/f Noise in General, *Solid-State Electron.*, 34, 1–21, 1991.
19. T. Musha and M. Tacano, Dynamics of Energy Partition among Coupled Harmonic Oscillators in Equilibrium, *Physica A*, 346, 339–346, 2005.
20. R. P. Jindal and A. van der Ziel, Phonon Fluctuation Model for Flicker Noise in Elemental Semiconductors, *J. Appl. Phys.*, 52, 2884–2888, 1981.
21. M. N. Mihaila, Phonon-Induced 1/f Noise in MOS Transistors, *Fluctuation Noise Lett.*, 4, L329–L343, 2004.
22. A. L. McWorther, *Semiconductor Surface Physics*, University of Pennsylvania Press, Philadelphia, USA, 1957.

23. G. Ghibaudo, O. Roux, Ch. Nguyen-Duc, F. Balestra, and J. Brini, Improved Analysis of Low-Frequency Noise in Field-Effect MOS Transistors, *Phys. Stat. Sol. A*, 124, 571–581, 1991.
24. K. K. Hung, P. K. Ko, C. Hu, and Y. C. Cheng, A Unified Model for the Flicker Noise in Metal-Oxide-Semiconductor Field-Effect Transistors, *IEEE Trans. Electron Dev.*, 37, 654–665, 1990.
25. G. Reimbold, Modified 1/f Trapping Noise Theory and Experiments in MOS Transistors Biased from Weak to Strong Inversion-Influence of Interface States, *IEEE Trans. Electron Dev.*, ED-31, 1190–1198, 1984.
26. X. Li, C. Barros, E. P. Vandamme, and L. K. J. Vandamme, Parameter Extraction and 1/f Noise in a Surface and a Bulk-Type, p-chAnnel LDD MOSFET, *Solid-State Electron.*, 37, 1853–1862, 1994.
27. F. N. Hooge and L. K. J. Vandamme, Lattice Scattering Causes 1/f Noise, *Phys. Lett. A*, 66, 315–316, 1978.
28. J. Rhayem, D. Rigaud, A. Eya'a, and M. Valenza, 1/f Noise in Metal-Oxide-Semiconductor Transistors Biased in Weak Inversion, *J. Appl. Phys.*, 89, 4192–4194, 2001.
29. T. G. M. Kleinpenning, On 1/f Noise and Random Telegraph Noise in Very Small Electronic Devices, *Physica B*, 164, 331–334, 1990.
30. N. V. Amarasinghe, Z. Çelik-Butler, and A. Keshavarz, Extraction of Oxide Trap Properties Using Temperature Dependence of Random Telegraph Signals in Submicron Metal-Oxide-Semiconductor Field-Effect Transistors, *J. Appl. Phys.*, 89, 5526–5532, 2001.
31. N. V. Amarsinghe and Z. C. Butler, Complex Random Telegraph Signals in 0.06μm² MDD nMOSFETs, *Solid State Electron.*, 44, 1013–1019, 2000.
32. W. Shockley and W. T. Read Jr., Statistics of the Recombinations of Holes and Electrons, *Phys. Rev.*, 87, 835–842, 1952.
33. J.-S. Lim, Strain Effects on Silicon CMOS Transistors: Threshold Voltage, Gate Tunneling Current, and 1/f Noise Characteristics, PhD thesis, University of Florida, 2007.
34. J. Lim, X. Yang, T. Nishida, and S. E. Thompson, Measurement of Conduction Band Deformation Constants Using Gate Direct Tunneling Current in n-Type Metal Oxide Semiconductor Field Effect Transistors under Mechanical Stress, *Appl. Phys. Lett.*, 89, 073509, 2006.
35. J. Chang, A. A. Abidi, and C. R. Viswanathan, Flicker Noise in CMOS Transistors from Subthreshold to Strong Inversion at Various Temperatures, *IEEE Trans. Electron Dev.*, 41, 1965–1971, 1994.
36. M. Ertürk, T. Xia, and W. F. Clark, Gate Voltage Dependence of MOSFET 1/f Noise Statistics, *IEEE Electron. Dev. Lett.*, 28, 812–814, 2007.
37. C. K. Maiti, L. K. Bera, S. S. Dey, D. K. Nayak, and N. B. Chakrabarti, Hole Mobility Enhancement in Strained-Si p-MOSFETs under High Vertical Field, *Solid-State Electron.*, 41, 1863–1869, 1997.
38. L. K. Bera, Studies on Applications of Strained-Si for Heterostructure Field-Effect Transistors (HFETs), PhD thesis, Indian Institute of Technology, Kharagpur, 1998.
39. van der Ziel A, Flicker Noise in Electronic Devices, *Adv. Electronics Electron Phys.*, 49, 225–297, 1979.

40. N. Singh, A. Agarwal, L. K. Bera, R. Kumar, G. Q. Lo, B. Narayanan and D. L. Kwong, High-Performance Fully Depleted Silicon Nanowire (Diameter ≤ 5 nm) Gate-All-Around CMOS Devices, *IEEE Electron Dev. Lett.*, 27, 383–386, 2006.

41. N. Singh, K. D. Buddharaju, S. K. Manhas, A. Agarwal, S. C. Rustagi, G. Q. Lo, N. Balasubramanian, and D.-L. Kwong, Si, SiGe Nanowire Devices by Top–Down Technology and Their Applications, *IEEE Trans. Electron Dev.*, 55, 3107–3118, 2008.

42. G. J. Zhang, G. Zhang, J. H. Chua, R. E. Chee, E. H. Wong, A. Agarwal, and K. D. Buddharaju, DNA Sensing by Silicon Nanowire: Charge Layer Distance Dependence, *Nano Lett.*, 8, 1066–1070, 2008.

43. C. Wei, Y.-Z. Xiong, X. Zhou, N. Singh, S. C. Rustagi, G. Q. Lo, and D.-L. Kwong, Investigation of Low-Frequency Noise in Silicon Nanowire MOSFETs in the Subthreshold Region, *IEEE Electron Dev. Lett.*, 30, 668–671, 2009.

44. D. Jimenez, B. Iniguez, J. Sune, L. F. Marsal, J. Pallares, and J. Roig, Continuous Analytical I-V Model for Surrounding-Gate MOSFETs, *IEEE Electron Dev. Lett.*, 25, 571–573, 2004.

45. M. V. Haartman and M. Ostling, *Low-Frequency Noise in Advanced MOS Devices*, Springer-Verlag, Dordrecht, Netherlands, 2007.

46. N. Siabi-Shahrivar, W. Redman-White, P. Ashburn, and H. A. Kemhadjian, Reduction of 1/f Noise in Polysilicon Emitter Bipolar Transistors, *Solid-State Electron.*, 38, 389–400, 1995.

47. T. G. M. Kleinpenning, Location of Low-Frequency Noise Sources in Submicrometer Bipolar Transistors, *IEEE Trans. Electron Dev.*, ED-39, 1501–1506, 1992.

48. M. V. Haartman, M. Sanden, G. Bosman, and M. Ostling, Random Telegraph Signal Noise in SiGe Heterojunction Bipolar Transistors, *J. Appl. Phys.*, 92, 4414–4421, 2002.

49. K. Kandiah, M. O. Deighton, and F. B. Whiting, A Physical Model for Random Telegraph Signal Currents in Semiconductor Devices, *J. Appl. Phys.*, 66, 937–948, 1989.

50. M. J. Deen, S. L. Rumyantsev, and M. Schroter, On the Origin of 1/f Noise in Polysilicon Emitter Bipolar Transistors, *J. Appl. Phys.*, 85, 1192–1195, 1999.

51. D. Knoll, B. Heinemann, K.-E. Ehwald, H. Rücker, B. Tillack, W. Winkler, and P. Schley, BiCMOS Integration of SiGe: C Heterojunction Bipolar Transistors, *Proc. IEEE BCTM 2002*, 162–166, 2002.

52. J. D. Cressler and G. Niu, *Silicon Germanium Heterojunction Bipolar Transistors*, Artech House, Norwood, Massachusetts, USA, 2003.

7

Technology CAD of Strain-Engineered MOSFETs

Technology computer-aided design (TCAD) refers to the use of computer simulation to model semiconductor processing and device operation. TCAD provides insight into the fundamental physical phenomena that ultimately impact performance and yield. Process variability is becoming more and more design for manufacturing challenges for designers as new process steps are added for advanced technology nodes. By integrating TCAD-derived models with physical design tools, the designer can focus on optimising variation awareness for increased performance, productivity, and predictability. From the 90 nm node onward, issues with manufacturability and yield have forced the electronic design automation (EDA) industry and manufacturing to move closer together. In particular, process and device information that affect functionality and yield need to be incorporated into the design flow, addressing more comprehensively issues of design for manufacturing (DFM) and yield (DFY). For true DFM and DFY, it is necessary to include process variability in the design process. From the 65 nm node, the variability has increased significantly further as a result of feature scaling and the introduction of new materials and innovative techniques such as strain engineering. The information needed by designers includes layout sensitivity as well as the effect of process variability on the electrical characteristics of devices and interconnects.

The strength of TCAD lies in the accurate prediction of device and interconnects variability due to layout as well as random variations in the process. Variability information can then be incorporated into the design tools through appropriate statistical compact models. Ultimately, this will lead to an improved design flow that addresses manufacturability issues in a comprehensive way. Beyond traditional process/device modelling and optimisation TCAD provides unique advantages in manufacturing. In a manufacturing for design approach, TCAD can be used for global advanced process control, helping the manufacturing engineer optimise parametric product yield for specific designs without having to understand the details of the design itself.

7.1 TCAD Calibration

TCAD acts as a bridge between the process and the device engineer. TCAD calibration refers to the process of selecting appropriate models and adjusting the model parameters so that the response of the physical model can predict the measured values. It may be noted that even the measured data being used for calibration (e.g., SIMS or I-V) may have experimental errors that cannot be controlled or estimated. If the goal is to predict a new technology that is being developed and is changing, TCAD calibration will be essentially a dynamic process. TCAD has technical limitations that must be addressed to achieve the full potential of TCAD in semiconductor technology development and manufacturing. Aside from the difficulties intrinsic to TCAD calibration, such as measurement errors and physical understanding, some problems arise from the way one develops and uses the models. As such, for industrial usage of TCAD for predictive process and device simulation, special care needs to be taken. The knowledge of the technical limitations of TCAD is crucial to set realistic goals and expectations from TCAD. The major technical limitation of TCAD, such as accuracy and predictability, must be addressed by a proper calibration of process and device models. During simulator (either process or device) development physical models are implemented in TCAD tools. For advanced device simulators, special emphasis on emerging topics, like quantum mechanical confinement, tunneling, and discrete dopant effects, is needed. For example, limitations of drift diffusion and hydrodynamic models in nanoscale device simulation are well known.

The critical issues that require special attention are physical model calibration, selection of effective physical models, numerical aspects, grid generation, and so on. The properly characterised process and device models can be effectively applied to develop fabrication process technology that can significantly reduce the development cycle time and cost. The general philosophy and the step-by-step procedure of numerical model calibration for predictive application of TCAD in technology and device design have been presented in [1]. In this approach, the entire product development cycle is divided into three phases: the generation of the initial process recipe, the optimisation of the process technology, and the evaluation of process manufacturability. Although currently the device simulation has a relatively strong basis, the new phenomena are of greater importance in the deep submicron devices, and the device performance predictions are inevitably linked to process TCAD. The role of benchmark standards and calibration testers in verifying the fundamental accuracy of the device simulators themselves is a challenge for the research community.

The elusive goal of TCAD is to achieve predictability of the final device characteristics, based on actual process conditions rather than idealised processes. Sensitivity studies must be included to isolate the important parameters.

Validation is considered to be more difficult than model or code development. Results are compared after prediction. For example, a measured 2D dopant profile can be compared with SUPREM predictions, and both can be used as inputs to a device simulator for predictions to assess comparisons for actual device results. An improvement in predictability will also provide greater confidence in reliability assessments and in process/structure configurations that improve robustness.

Since the speed of metal-oxide-semiconductor field-effect transistor (MOSFET) is intrinsically limited by the carrier transit time, the most obvious approach to improve the device speed is to reduce the gate length. However, the 2D effects due to the gate length scaling affect the threshold voltage and subthreshold slope and increase the off-state current. When the gate length is aggressively scaled, the gate begins to lose control over the channel and the parasitic conduction layers, resulting in saturation, possibly a reduction of transconductance, and an increase of drain conductance. It is now well recognised that under the sub-100 nm regime, conventional MOSFET scaling concepts will be confronted with physical limitations—the lattice constant of Si. The major problems of scaling down the conventional MOSFETs include (1) quantum mechanical tunneling through the thin gate oxide, from source to drain and from drain to body; (2) threshold voltage control induced by random doping effects; (3) short-channel effects and mobility degradation; and (4) process control of thin layer uniformity and accurate lithography and implantation.

To extend the lifetime of the Si-based complementary metal-oxide-semiconductor (CMOS) technology, devices with new structures or new materials need to be considered [2]. Several performance boosters have been proposed in the literature, and the International Technology Roadmap for Semiconductors suggests that one or more technology boosters may be required for devices beyond the 45 nm technology node in order to sustain the increase of intrinsic device speed. Process steps introduce compressive stress in the Si channel of a MOSFET. Through appropriate strain engineering, such as shallow trench isolation (STI), embedded SiGe under the source/drain region, and the cap layer or silicidation process (all of which have been incorporated into Intel's current 90 nm technology) performance enhancement can be achieved. In comparison with the process-induced strain, strained Si pseudomorphically grown on a relaxed SiGe layer is a commonly adopted approach to achieve tensile strain. Strained Si on a relaxed SiGe buffer for CMOS applications has been studied for more than 15 years. Progress has also been made toward the evolution of the strained Si MOS structure, such as the strained Si on SiGe-on-insulator (SGOI) and the strained Si-directly-on-insulator (SSDOI). With a highly strained Si channel or using a different orientation (110) substrate in p-MOSFETs, the performance match between the n- and p-MOSFETs for CMOS applications might be achieved.

In this chapter we present the results of our simulation study on the evolution of Si-based MOSFETs via the incorporation of new materials, for example, strained Si, to predict the future device performance and the scaling trends of strained Si/SiGe MOSFETs for radio frequency (RF) applications. We also propose to combine the advantages of the hybrid orientation technology (HOT) and process-induced local strain (PSS) engineering to improve CMOS device performance via mobility engineering. Mobility models are developed and used in predictive simulation. The work is based on simulation, using the commercial TCAD tools from Synopsys. Based on the calibration of the models used, performance predictions of scaled strained Si MOSFETs are made.

7.2 Simulation of Strain-Engineered MOSFETs

In the following, process-induced stress simulations are performed using the SProcess simulator with the help of a set of strain models for both the n- and p-MOSFETs. The basic equations used to compute the strain and stress (within the elastic limit) in a global equilibrium condition are given by

$$\sum_{k=x,y,z} \frac{\partial \sigma_{jk}(v)}{\partial x_k} = 0 \quad (\text{for } j = x, y, z) \tag{7.1}$$

Equation (7.1) is solved using the finite element method (FEM). The stress present in a MOSFET may be divided into two parts: (1) lattice mismatch stress and (2) intrinsic stress. Lattice mismatch stress occurs when two materials with different lattice constants expand or contract at different rates. If the lattice spacing of Si is L_{Si} and that for SiGe is L_{SiGe}, then the strain is given by

$$\varepsilon = \left(L_{SiGe} - L_{Si}\right)/\left(L_{Si}\right) \tag{7.2}$$

Intrinsic stress is generated due to several factors, such as deposition rate, thickness, and temperature. During the deposition process, thin films are either stretched (creating intrinsic tensile stress) or compressed (creating intrinsic compressive stress) to fit the substrate on which they are deposited.

The SProcess [3] tool is used to simulate and optimise a typical 45 nm process flow, including channel, halo, source/drain (S/D) engineering, oxidation, deposition, etching, and annealing for dopant activation. The stress history is calculated for the entire process flow. Rapid thermal annealing (RTA) of implanted profiles is not taken into account because, for such a rapid process, no noticeable stress relaxation has been observed. Strain sources are lattice mismatch (SiGe pocket) and intrinsic stress (compressive cap). A source/drain SiGe pocket is formed with 17% Ge at room temperature, and nitride cap layers are introduced after the critical doping steps, and therefore have negligible impact on the final doping distributions.

A three-stream diffusion model is used. Additionally, formation of point defect clusters and the three-phase segregation model accounting for the dose loss at the silicon/oxide interface are considered. For simulation of ion implantation, a 2D analytic integration with dual-Pearson (in silicon) and Pearson distribution functions (in other materials) is used. In process simulation, the Hobler model is used to simulate the damage profiles and amorphisation. For the simulation of Ge diffusion and redistribution in strained Si a model is developed that supports structures with various regions containing strained Si. The effect of change in material composition due to Ge diffusion and its effect on strain (for p-MOSFETs) are also incorporated. At each diffusion step, the stress evolution is computed, including oxidation steps based on the viscoelastic model. This model is also used to compute the stress effects on n-MOSFETs after the deposition of the highly tensile cap layer. Although successful demonstrations of both n- and p-type experimental strained Si MOSFETs have been made, little information is available on the performance of scaled strained Si MOSFETs with gate length less than 50 nm. Note that the strained Si layer thickness is kept unchanged since simulations show that variation of the strained Si layer thickness from 5 nm to 15 nm has negligible effects on the output characteristics. Figure 7.1

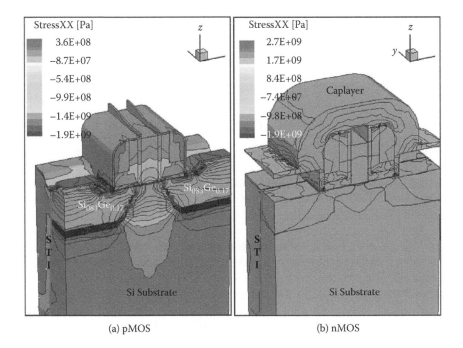

(a) pMOS   (b) nMOS

FIGURE 7.1
Three-dimensional device structure of 45 nm devices obtained from SProcess simulation: (a) p-MOSFET and (b) n-MOSFET.

TABLE 7.1

Major Technology Parameters Used for Process-Induced Strained CMOS Fabrication

Parameter	PSS-p-MOSFET	PSS-n-MOSFET
Channel implants	P, 370 KeV,	B, 300 KeV, 3.0e13 cm^{-2}
-Well	2.65e13 cm^{-2}	B, 120 KeV, 2.05e13 cm^{-2}
-V_{th} adjustment	P, 260 KeV,	B, 50 KeV, 1.1e13 cm^{-2}
	2.65e13 cm^{-2}	B, 25 KeV, 1.0e13 cm^{-2}
	P, 40 KeV, 1.0e13 cm^{-2}	
	B, 150 KeV, 1.0e13 cm^{-2}	
Poly-doping	BF2, 10 KeV,	P, 10 KeV, 2.1e15 cm^{-2}
	2.1e15 cm^{-2}	
Halo implants	As, 20 KeV, 5.0e13 cm^{-2}	B, 10 KeV, 6.0e13 cm^{-2}
Source/drain extension (SDE)	B, 1.0 KeV, 1.0e14 cm^{-2}	As, 5.0 KeV, 8.0e14 cm^{-2}
Deep source/drain (highly doped drain (HDD))	B, 5.0 KeV, 1.0e15 cm^{-2}	P, 15 KeV, 1.5e15 cm^{-2}
Final RTA	1025°C, 1.0 s	1025°C, 1.0 s

shows the stress distribution of the 3D device structure obtained from SProcess simulation.

The structure generated by SProcess is then simulated using SDevice [4]. The simulated device performance includes DC electrostatic behaviour with strain-induced mobility enhancement and the impact of rapid thermal annealing (RTA) on device performance. Table 7.1 shows the major process parameters used in simulation. A hydrodynamic transport model was used for all simulations. In addition, a strain-specific model is also used to capture the influence of stress on carrier transport.

7.2.1 Strain-Engineered p-MOSFETs

Embedded SiGe source/drain regions are used to incorporate the compressive stress in p-MOSFETs. SiGe pockets are introduced in source/drain regions by selective epitaxy. The strain calculation for this layer includes the compression due to Ge incorporation. The $Si_{0.83}Ge_{0.17}$ pockets induce uniaxial compressive stresses in different areas of the structure, including the channel. This may be seen in the stress distribution after S/D anneals, as presented in Figure 7.1(a). The stresses are computed as a function of the lattice mismatch between unstrained Si and SiGe. Again, it is clear that the stresses induced by the SiGe pockets significantly alter the stress distribution in the channel area. The strain in the channel may be altered by Ge mole fraction in the pocket, stressor depth, stressor height, and gate length. The effects of these parameters on stress developed are discussed below. The stress (both x component ε_{xx} and y component ε_{yy}) along the channel from source to drain region for different stressor depths is shown in Figure 7.2.

It is observed that the stress is high at the centre of the channel. The x component of stress in the channel is more compressive for large

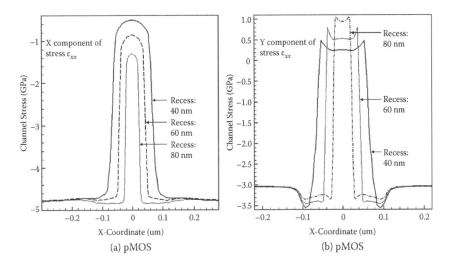

FIGURE 7.2
Profiles of the (a) lateral stress component ε_{xx} and (b) vertical stress component ε_{yy} in a transistor structure for different stressor depths where the gate length L is 45 nm.

stressor depths, and the y component of stress is more tensile. Variation of stress with stressor depths is shown in Figure 7.3. As the depth of the SiGe stressors is increased (all other scaling parameters are kept fixed), the average lateral compressive stress in the silicon channel is increased. Negative stress values indicate compressive stress. For short gate length, stress inside

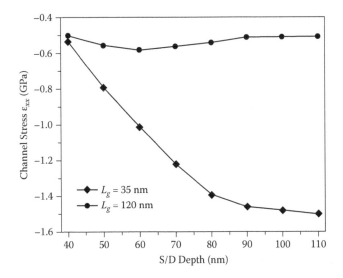

FIGURE 7.3
Stress boosting in p-MOSFETs by increasing SiGe depth for 35 and 120 nm gate length devices.

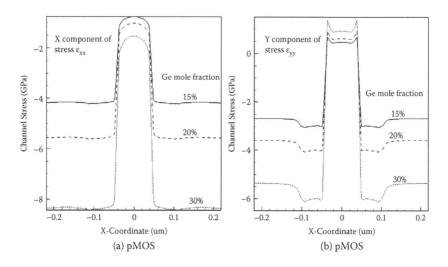

FIGURE 7.4

Effect of Ge concentrations for stress profile in p-MOSFETs of channel lengths 45 nm, and 40 nm stressor depth: (a) x component of stress ε_{xx} and (b) y component of stress ε_{yy}.

the channel is much higher than the longer one. Raising the SiGe source/drain depth to a certain extent transfers higher stress to the channel, thereby further improving the mobility of holes.

Although the stress induced in the channel due to Ge incorporation is insignificant for long-channel devices, for CMOS transistors with channel lengths in the nanometer range, the stress developed plays a significant role in determining the carrier mobility enhancement. Figure 7.4 shows that the stress component of (a) ε_{xx} and (b) ε_{yy} for a given gate length. As the Ge mole fraction (or the lattice mismatch) between the stressor and the channel increases, the magnitudes of both ε_{xx} and ε_{yy} are found to increase linearly.

The average strains for Ge mole fractions of 15, 20, and 30% are computed, and as expected, one needs to use a higher Ge concentration to obtain a higher strain, as shown in Figure 7.5, where a higher channel stress is obtained when the Ge in S/D is increased to 20% or 30% for the same gate length.

The switching speed of a transistor can be increased primarily by physical gate length downscaling. The channel stress increases as the gate length is scaled. The variation of stress with gate length for different recess depths is shown in Figure 7.6. Simulation shows that decreasing the gate length assists in boosting the stress transferred into the device channel.

7.2.2 Strain-Engineered n-MOSFETs

A highly tensile nitride cap layer is used to improve the performances of n-MOSFETs. The nitride film transfers the stress to the channel because

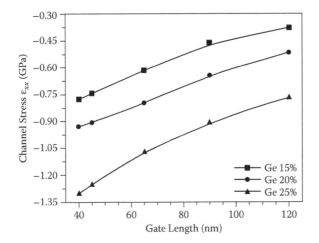

FIGURE 7.5
Effect of varying gate length in p-MOSFETs for different Ge concentrations.

an edge force is developed as the film grows over the spacer and gate. The channel stress developed depends on various parameters, such as nitride thickness, poly-thickness, spacer width, gate length, and S/D opening. Figure 7.1(b) shows the device simulated with a highly tensile cap layer. It shows that stresses in the cap layer lead to tensile stress in the channel area of n-MOSFETs.

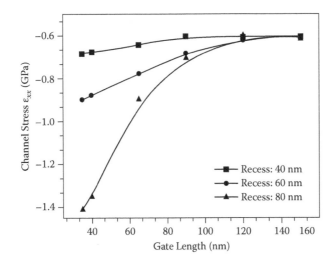

FIGURE 7.6
Channel stress in p-MOSFETs as a function of gate length for 45, 60, and 80 nm stressor depths.

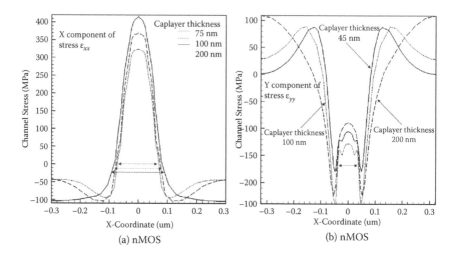

FIGURE 7.7
Effect of increasing silicon-nitride thickness on channel: (a) lateral stress ε_{xx} components and (b) vertical stress components ε_{yy}.

Stress distribution from source to drain regions of 45 nm n-MOSFETs for different cap layer thicknesses are shown in Figure 7.7. As shown in Figure 7.7(a), the cap layer thickness increase results in higher tensile ε_{xx} stress in the channel region. Tensile stress is larger at the centre of the channel. Figure 7.7(b) shows the ε_{yy} component stress distribution. The variation of channel stress, both ε_{xx} and ε_{yy}, with nitride thickness is shown in Figure 7.8. Figure 7.9 shows

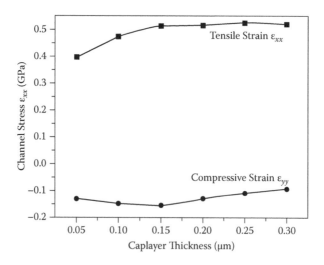

FIGURE 7.8
Effect of increasing silicon-nitride thickness on channel stress components in n-MOSFETs.

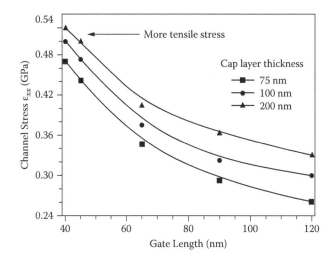

FIGURE 7.9
Effect of scaling gate length on n-MOSFETs channel stress for different cap layer thicknesses.

that channel stress increases as the gate length is scaled down since the channel is in closer proximity to the tensile capping silicon-nitride layer for smaller critical dimensions. For a fixed gate length, stress is more for higher cap layer thickness. However, it is difficult to strain long-channel devices, compared to their short-channel counterparts, using the tensile nitride capping layer, which is an important consideration for circuit designers while designing for optimum circuit performance.

7.3 DC Performance

The MOSFET structure used in simulation was chosen from reference [5], as reliable experimental data are available for benchmarking the predictive simulation results. Briefly, the MOSFETs have a gate length of 45 nm with 1.2 nm gate oxide. Experimental data were reported for two different drain biases, and the measured drain current vs. gate voltage is shown (Figure 7.10), along with our simulation results. A good agreement is observed showing the prediction capability of TCAD simulation.

Figure 7.11 shows the $I_{ds}(I_d)$-V_{ds} characteristics of the 45 nm MOSFETs with and without strained Si channel. For the n-MOSFETs, the simulated results indicate an approximately 23% increase in drain current at $V_{ds} = V_{gs} = 1.2$ V due to an enhancement in electron mobility as a result of the strain in the channel. An empirical relationship between the strain components and the

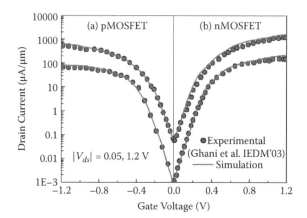

FIGURE 7.10
Calibration of simulated output characteristics with experimental device.

linear drain current has been reported in [6]. The change in linear drain current for p-MOSFETs may be expressed as [7]

$$\frac{I_{dlinStSi} - I_{dlinSi}}{I_{dlinSi}} = \frac{\Delta I_{dlin}}{I_{dlin}} = a_x\varepsilon_{xx} + a_y\varepsilon_{yy} + a_z\varepsilon_{zz} \quad (7.3)$$

where a_x, a_y, and a_z are the strain sensitivity coefficients with respect to the x, y, and z strain components, respectively. Since $\Delta I_{dlin}/I_{dlin} \approx \Delta\mu/\mu$, the mobility

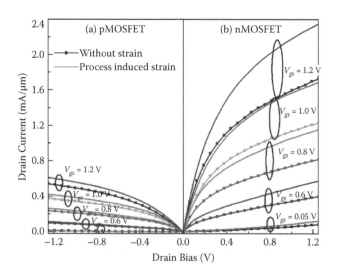

FIGURE 7.11
Simulated I_d-V_d characteristics with different gate biases.

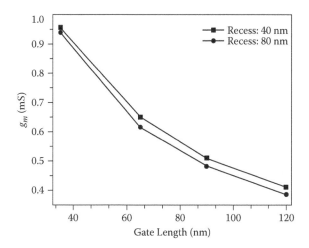

FIGURE 7.12
Transconductance vs. gate length for 40 and 80 nm recesses in p-MOSFETs.

enhancement is approximately the same as the linear drain current enhancement. Using Equation (7.3), we have computed the electron and hole mobility enhancement factor due to tensile and compressive longitudinal stress for strain-engineered n- or p-MOSFETs, respectively. For the PSS p- and n-MOSFETs, hole and electron mobility enhancement factors have been found to be ~1.5× and ~1.8× that of bulk Si, which is also consistent with our simulation results. The resulting simulation demonstrates an approximately 17% enhancement of drain current with respect to the conventional silicon p-MOSFET. The impact of a strained Si channel on device performance is evaluated in Sentaurus Device by enhancing the mobility by an amount consistent with that which can be realised in practise with embedded SiGe S/D layers for p-MOSFETs and a highly tensile silicon-nitride cap layer for n-MOSFETs.

The variation of g_m with gate length (for p-MOSFET) for different recesses is shown in Figure 7.12. It is seen that for the same gate length, improvement in g_m is mainly caused by the process-induced mechanical stress.

Thermal annealing after the SiGe S/D pocket formation may slightly alter the doping profile and effective gate length compared to a reference device. As such, the absolute g_m improvement, as shown in Figure 7.13, should be interpreted with care.

Simulated subthreshold slope (SS) vs. channel stress is shown in Figure 7.14 for the case where the stress is modulated by a SiGe recess. Figure 7.15 shows the g_m variation vs. cap layer thickness for different gate lengths in n-MOSFETs.

Figure 7.16 shows the transconductance improvement for 45 nm gate length transistors over the reference transistor with the same gate length. A 24% higher g_m than for reference transistors is obtained. The subthreshold slope also depends on nitride film thickness and is shown in Figure 7.17.

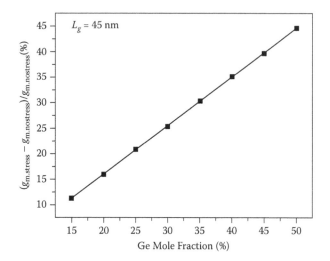

FIGURE 7.13

Transconductance enhancements for $L = 45$ nm channel length transistors, as a function of the germanium mole fraction in the source/drain (S/D) region of the p-MOSFETs.

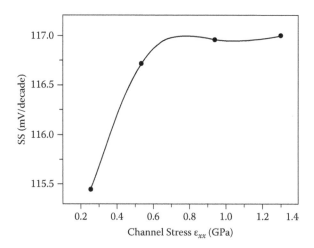

FIGURE 7.14

Simulated subthreshold voltage swing ($V_{gs} = V_{ds} = 1.2$V) of p-MOSFETs.

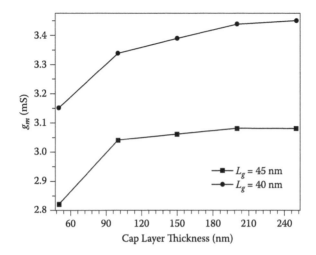

FIGURE 7.15
Transconductance vs. cap layer thickness for gate lengths 45 and 40 nm in n-MOSFETs.

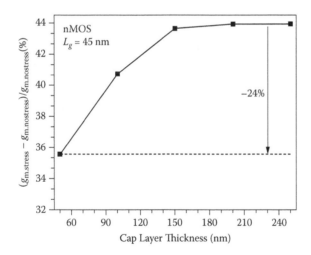

FIGURE 7.16
Maximum transconductance improvement (V_{ds} = 1.2 V) of n-MOSFETs over reference transistors (no stress) with the same mask with a gate length of 45 nm. The arrow indicates the relative reduction of the g_m improvement with respect to reference devices for transistors with a 230 nm cap layer thickness.

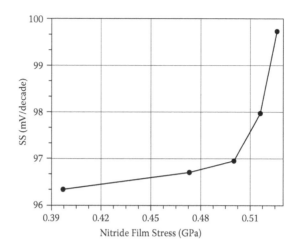

FIGURE 7.17
Simulated subthreshold voltage swing ($V_{gs} = V_{ds} = 1.2V$) of n-MOSFETs.

7.4 AC Performance

Since scaled p- and n-MOSFETs may find applications in the RF regime, it is necessary to develop some figures of merit to assess the device performance. Using SDevice transient simulation, based on the mobility models described before, small-signal analysis has been performed to extract RF parameters for process-induced strained Si MOSFETs.

An AC simulation is performed at equidistant bias points for small-signal AC analysis at various frequencies, with the gate as the input port, the drain as the output port, and the source and substrate grounded. The resulting small-signal admittance and capacitance parameters are then used to extract a RF figure of merit, such as the cutoff frequency (f_T). The bias dependence of f_T is shown in Figure 7.18. We extracted f_T using the following expression, in which the current gain, $|h_{21}|$, is converted from the s parameters using the transformation:

$$|h_{21}| = \left| \frac{-s_{21}}{(1 - s_{11})(1 + s_{22}) + s_{12}s_{21}} \right| \qquad (7.4)$$

7.5 Hybrid Orientation Technology for Strain-Engineered MOSFETs

In this section, we present the technology computer-aided design (TCAD) and simulation results for both the stress engineered n- and p-MOSFETs on <100> and <110> hybrid orientation substrates, respectively. The source

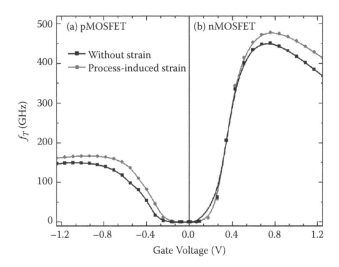

FIGURE 7.18
Cutoff frequency (f_T) as a function of gate bias for process-induced strained Si p- and n-MOSFETs.

and drain regions are idealised by a short box doped to 1×10^{20} cm^{-3} and present negligible series resistance. The value of the gate resistance (R_{gate}) is added via postprocessing to the TCAD simulations. Process-induced strained p-MOSFET with 45 nm gate length was simulated for different surface orientation. Figure 7.19(a) compares the drain current against the gate voltage characteristics for the devices with compressive stress. The corresponding 2D device simulation shows a slightly higher drive current due to stress. An improvement in drive current in the <110> direction is observed over the drive current in the <100> direction under the longitudinal uniaxial compressive stress and is of much significance. For comparison, 45 nm gate length process-induced strained channel n-MOSFETs with different surface orientations were also simulated. Figure 7.19(b) compares the drain current against the gate voltage characteristics for the devices with a highly tensile cap layer and a relaxed capping layer. The corresponding 2D simulation shows a higher current (of the order of 14–15%) gain due to stress. Figure 7.20 compares the drain current against the drain voltage characteristics. In case of n-MOSFETs, devices with a highly tensile cap layer show a slightly higher current gain due to stress (~15%) than the device with a relaxed capping layer.

It also shows a drive current improvement in the <100> direction, more than in the <110> under the longitudinal uniaxial tensile stress, and it is of much significance to HOT. Also, p-MOSFETs with compressive stress show higher drain current than bulk Si MOSFETs. A higher drive current in the <110> direction than in the <100> direction is observed under the longitudinal uniaxial compressive stress. Figure 7.21 shows a comparison of threshold voltage for both the HOT MOSFETs and PSS MOSFETs.

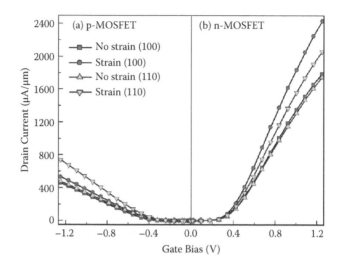

FIGURE 7.19
Comparisons of I_{ds}-V_{gs} characteristics in PSS (a) p- and (b) n-MOSFETs.

The variation of f_T with V_{gs} for PSS p- and n-MOSFET is shown in Figure 7.22(a) and (b), respectively. It is found that the f_T is higher in the <110> direction than in the <100> direction with stress for PSS p-MOSFETs. For PSS n-MOSFETs, f_T is higher in the <100> direction than the <110> direction with stress and without stress. These results indicate the advantages of strain-dependent mobility enhancement along with hybrid orientation toward a

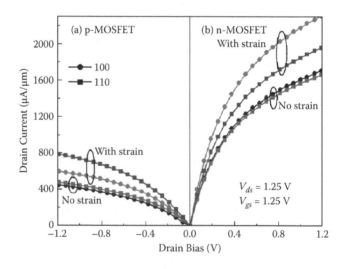

FIGURE 7.20
Comparisons of I_{ds}-V_{ds} characteristics in PSS (a) p- and (b) n-MOSFETs.

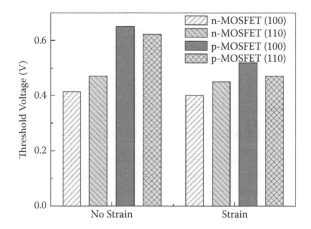

FIGURE 7.21
Comparison of threshold voltage in PSS p-MOSFETs and n-MOSFETs.

high-speed strain-engineered CMOS device design. Since f_T depends on both the C_{gs} and g_m, it is important to clarify the impact. The improvements observed in devices are driven primarily by stress-induced enhancement of g_m as C_{gs} is almost identical for devices with different stress and orientation configurations.

From simulation, f_T values of 510.65 and 478.2 GHz for uniaxial <100> and uniaxial <110> n-MOSFETs and 227.9 and 245.59 GHz for uniaxial <100> and

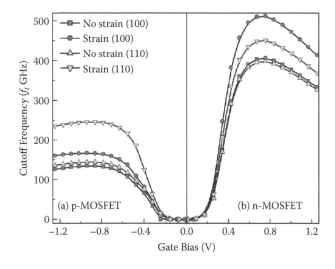

FIGURE 7.22
Variation of f_T with gate bias in process-induced strained Si (a) p-MOSFETs and (b) n-MOSFETs.

uniaxial <110> p-MOSFETs, respectively, have been obtained. The highest f_T values for uniaxial strain n-MOSFETs in the <100> direction and p-MOSFETs in the <110> are due to higher orientation-dependent mobility enhancement.

7.6 Simulation of Embedded SiGe MOSFETs

A new strained silicon concept that utilises elastic relaxation of a buried compressive SiGe layer to induce tensile strain in the channel has been reported [8]. The Rev. e-SiGe (reverse e-SiGe) technique has been shown to be effective, inducing a level of stress comparable to or exceeding conventional strained silicon techniques, and it is shown to be scalable down to a gate length of 10 nm. Donaton et al. [9] have demonstrated a substantial drive current enhancement in sub-100 nm n-MOSFETs and basic simulations of the influence of the device parameters on the channel stress. These results were quite promising. In the following, we present an extensive simulation study of a MOSFET with embedded SiGe. Simulations were performed to calculate the channel stress for device structures. For fabrication, a standard CMOS process is completed through the STI step, and then the n-MOSFET active areas are etched to create a small recess, and thin compressed SiGe and relaxed silicon layers are epitaxially grown on the active areas. The SiGe layer is compressively strained because its lattice constant is larger than the lattice constant of silicon. Then, the process is continued through a standard gate stack process, including gate oxide growth, gate and silicon-nitride cap deposition, gate etch, extension, halo implant, and spacer definition. The source/drain areas are then etched. This is the most important step in the process, as it creates a lateral free surface allowing the compressed buried SiGe layer to elastically expand, reducing the compressive stress in the SiGe and inducing tensile stress in the silicon above. Silicon is then regrown in the recessed source/drain areas, and the CMOS fabrication process is continued to completion.

ATHENA [10] simulation of stress in a MOSFET structure with an embedded SiGe layer is discussed below. The simulated process includes epitaxial growth of thin compressed SiGe and relaxed silicon layers. The standard gate stacks are then emulated by gate oxide and poly-deposition and oxide spacer formation. All important geometrical characteristics of the test structure, including thicknesses of different layers, spacer width, gate length, etc., are parameterised. This allows investigation of effects of the parameter variations on important device characteristics. The simulation parameter set used approximately corresponds to those reported for experimental MOSFETs [8]. The most important step of this simulation run is etching of the source/drain areas because it creates free surfaces on the sides of the buried SiGe layer. This step results in elastic expansion of the buried layer, reducing the compressive stress inside the layer and generating tensile stress in the silicon

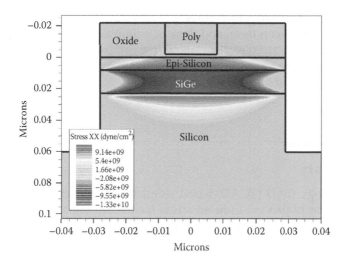

FIGURE 7.23
Two-dimensional contours of the x component of stress in the final structure.

above, i.e., under the gate. This enhanced tensile stress affects carrier transport and effectively improves device characteristics. The stress in the whole structure are calculated before and after the S/D etch. Figure 7.23 displays the 2D contours of the x component of stress in the final structure. Figure 7.24 compares stress profiles through the centre of the gate before and after the

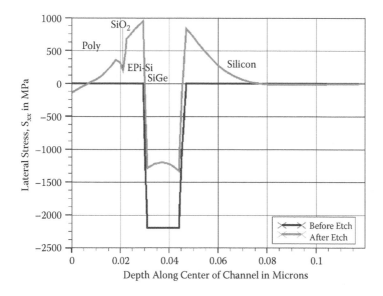

FIGURE 7.24
Stress profile through centre of gate before and after etch for structure.

S/D etch step. As shown in Figures 7.23 and 7.24, prior to etch, there is no stress in the top silicon layer, and there is uniform compressive stress in the SiGe layer. The etch step creates a lateral free surface, allowing the SiGe layer to expand, reducing its compressive stress and transferring tensile stress into the overlying silicon layer.

7.7 Summary

The technology CAD (TCAD)-based simulation approach, which includes the modelling of process-induced stress for the scaling studies of strained Si MOSFETs, is discussed in detail. Process and device simulations are performed to verify the reported experimental results for p-MOSFETs with embedded SiGe pockets and n-MOSFETs with tensile nitride capping layers. It is shown that p-MOSFET performance can be improved significantly when S/D regions are filled with SiGe pockets. For a given technology node, several options exist to increase the channel stress. For p-MOSFETs, increase of recess depth, Ge concentration, and decrease in channel length incorporate higher stress. For n-MOSFETs, increase of cap layer thickness and decrease of gate length would generate higher stress. Very high f_T values have been obtained for process-induced strained Si MOSFETs and need to be verified experimentally.

DC/AC performance of process-induced strained Si n- and p-MOSFETs in hybrid orientation technology has been studied using technology CAD tools that properly account for the physical mechanisms, such as orientation-dependent and process-induced strain-dependent mobility models. We have studied the effects of mobility enhancement, induced by surface orientation change, and also process-induced strain, simultaneously, on the RF performance of CMOS devices. Peak f_T values of about 524 and 239 GHz are predicted for n- and p-MOSFETs, respectively, in hybrid orientation technology involving process-induced strain. Our predictive simulation results have shown the superiority of hybrid orientation technology.

Review Questions

1. What is technology CAD?
2. What is process simulation?
3. What is device simulation?
4. What is the importance of TCAD in DFM?

5. What is hybrid orientation technology?

6. Mobility is dependent on substrate orientation. (True/False)

7. What are the sources of process variations?

8. Discuss how process variations affect the device performance. Give an example.

9. Discuss how process information can be included at the design stage.

10. What is TCAD calibration? Why it is important?

References

1. S. K. Saha, Modeling Process Variability in Scaled CMOS Technology, *IEEE Design Test Comput.*, 8–16, 2010.
2. C. K. Maiti, S. Chattopadhyay, and L. K. Bera, *Strained-Si Heterostructure Field Effect Devices*, CRC Press, Boca Raton, FL, 2007.
3. Synopsys, Inc., *Sentaurus Process User Manual*, version A-2007.12, Mountain View, CA, March 2008.
4. Synopsys, Inc., *Sentaurus Device User Manual*, version A-2007.12, Mountain View, CA, March 2008.
5. T. Ghani, M. Armstrong, C. Auth, M. Bost, P. Charvat, G. Glass, T. Hoffmann, K. Johnson, C. Kenyon, J. Klaus, B. McIntyre, K. Mistry, A. Murthy, J. Sandford, M. Silberstein, S. Sivakumar, P. Smith, K. Zawadzki, S. Thompson, and M. Bohr, A 90nm High Volume Manufacturing Logic Technology Featuring Novel 45nm Gate Length Strained Silicon CMOS Transistors, *IEEE IEDM Tech. Dig.*, 978–980, 2003.
6. T. K. Maiti, S. S. Mahato, S. K. Sarkar, and C. K. Maiti, Performance Enhancement of p-MOSFETs with Embedded SiGe Source/Drain on Hybrid Orientation Substrates, *Proc. Int. Conf. Ultimate Integration Silicon (ULIS 2007)*, Belgium, pp. 26–30, 2007.
7. Y.-C. Yeo, J. Sun, and E. H. Ong, Strained Channel Transistor Using Strain Field Induced by Source and Drain Stressors, *Proc. Mat. Res. Soc. Symp.*, 809, B10.4.1–B10.4.6, 2004.
8. J. G. Fiorenza, J.-S. Park, and A. Lochtefeld, Detailed Simulation Study of a Reverse Embedded-SiGe Strained-Silicon MOSFET, *IEEE Trans. Electron Dev.*, 55, 640–648, 2008.
9. R. A. Donaton, D. Chidambarrao, J. Johnson, P. Chang, Y. Liu, W. K. Henson, J. Holt, X. Li, J. Li, A. Domenicucci, A. Madan, K. Rim, and C. Wann, Design and Fabrication of MOSFETs with a Reverse Embedded SiGe (Rev. e-SiGe) Structure, *IEEE IEDM Tech. Dig.*, 465–468, 2006.
10. Silvaco International, *ATHENA/ATLAS User's Manual*, Santa Clara, CA, USA, 2009.

8

Reliability and Degradation of Strain-Engineered MOSFETs

Due to metal-oxide-semiconductor field-effect transistor (MOSFET) down-scaling, gate electric field increases. With the increase in electric field, an increase in chip operating temperature takes place, which is a serious reliability concern in silicon-integrated circuits. Moreover, the recent introduction of high-k gate dielectrics, metal gate materials, high-mobility channels, and new 3D device architectures has created a need for clear understanding of the reliability issues not only to negative bias temperature instability (NBTI) but also to positive bias temperature instability (PBTI). Transistors for three different types of logic are specified in the International Technology Roadmap for Semiconductors (ITRS): high performance (HP), low standby power (LSTP), and low operating power (LOP). To meet the performance and leakage current targets, key technology innovations i.e., high-k gate dielectrics and metal gate electrodes, ultra-thin body fully depleted silicon-on-insulator (SOI) MOSFETs, and multiple-gate MOSFETs, have been introduced in current complementary metal-oxide-semiconductor (CMOS) processing. New generation devices are taking advantage of the properties of high-k gate dielectrics, carrier mobility enhancement techniques, and new 3D architectures. In particular, the Ni fully silicided gate electrodes, techniques for local strain introduction, such as SiGe in source/drain and contact etch stop layers (CESLs), and the 3D FinFET technology have led to different types of reliability issues. This chapter will focus on the bias temperature instability (BTI) phenomenon in relation to technology scaling. NBTI and hot-carrier injection (HCI) degradation and their impact on strain-engineered MOSFETs are discussed.

Negative bias temperature instability (NBTI) was first reported in 1966 [1]. NBTI is one of the most important threats to p-MOSFETs in VLSI circuits. The electrical stress ($V_{gs} < 0$) on the transistor generates traps at the Si/SiO$_2$ interface. These defect sites increase the threshold voltage, reduce channel mobility of the MOSFETs, induce parasitic capacitances, and degrade the device performances [2]. NBTI is characterised by an increase in the absolute threshold voltage, and a degradation of the mobility, drain current, and transconductance under the influence of an applied gate voltage stress at elevated temperature. It is generally attributed to the creation of interface traps and oxide charge, although it is not clear which mechanism is dominant. The mechanism is ascribed to breaking of Si-H bonds at the

SiO$_2$/Si substrate interface by a combination of electric field, temperature, and holes, resulting in dangling bonds or interface traps at that interface. Various NBTI models have been proposed in the literature, of which the reaction-diffusion (R-D) model and the disorder-control-kinetics (DCK) are the most prevalent. In the R-D model, interface traps are generated at the SiO$_2$/Si interface (reaction) with a linear dependence on stress time. Hydrogen is released during the reaction phase and in the subsequent diffusion phase; the hydrogen diffuses away from the interface into the oxide. In this chapter we focus on distinguishing the interfacial defects intrinsic to the presence of strain from extrinsic defects associated with specific processing conditions and device geometry, which can be alleviated by processing optimisation. Some extrinsic factors, such as germanium out-diffusion from a virtual Si$_{1-x}$Ge$_x$ substrate and hydrogen diffusion from hydrogen-rich CESL liners, can become a reason for reliability deterioration of devices with process-induced strain.

Although the negative bias temperature instability in p-MOSFETs has attracted a lot of attention, for the positive bias temperature instability in n-MOSFETs relatively little has been published. The main reason for this is due to negligible PBTI effects observed in n-MOSFETs with SiO$_2$ or SiON dielectrics. However, with the introduction of high-k gate dielectrics, it can be an important reliability issue. PBTI is mostly described as electron trapping in native traps in the high-k layer. Intrinsic defects at the Si/SiO$_2$ interface are important in the operation of MOSFET devices. Unsaturated dangling bonds occur at the interface between the Si substrate and the oxide. Dangling bonds are formed at the interfaces between two materials with different lattice constants as a result of mismatch. The appearance of the dangling bonds depends on the crystallographic interface orientation.

MOSFETs gate oxide quality degrades during device operation and cannot retain its original condition. One general degradation type is defect generation in the oxide bulk or at the Si/SiO$_2$ interface over time. These defects increase leakage current through the gate dielectric, change transistor metrics such as the threshold voltage, or result in the device failure due to oxide breakdown. A very important step in the evaluation of these devices is the assessment of their reliability performance. The bias temperature instabilities (BTIs) are of utmost importance for determining the lifetime of the devices. Most NBTI measurements are made by stressing the device, then measuring the threshold voltage, interface trap density, drain current, transconductance, and other device parameters. There is usually a time delay between stress and characterisation, and the delay time is generally not mentioned in the literature. But the time delay is very important, as it may take seconds to generate NBTI damage, but the recovery is much faster, generally within a few microseconds. However, it is not clear if the NBTI damage recovers completely. It has been proposed that the damage consists of two degradation components: a permanent component that remains after stress removal and a reversible component that recovers.

Hot-carrier injection (HCI), although almost alleviated in current generation n-MOSFETs, is another mechanism that can also create defects at the Si/SiO$_2$ interface near the drain edge as well as in the oxide bulk [3]. Similar to NBTI, the traps shift the device parameters and degrade the device performance. The damage is due to carrier heating in the high electric field near the drain side of the MOSFET, resulting in impact ionisation and subsequent degradation. Historically, HCI has been more significant in p-MOSFETs because electrons have higher mobilities (due to lower effective mass) than holes, and thus can gain higher energy from the channel electric field. HCI has a faster rate of degradation than NBTI. HCI occurs during the low-to-high transition of the gate of an n-MOSFET; therefore, the degradation increases for high switching activity or higher frequency of operation.

In integrated circuits, MOSFETs operate under various stress conditions at different times, and are therefore exposed to different degradation types. For instance, in a CMOS inverter, the fundamental building block of the digital integrated circuits (ICs), both the n-MOSFET and the p-MOSFET, are tied to the same input voltage [4]. When the input signal is low ($\approx$ 0 V), the p-MOSFET is under NBTI stress, and therefore degrades while the n-MOSFET is turned off. When the input is pulled to high (V_{DD}), the n-MOSFET goes through an impact ionisation condition and experiences HCI degradation. At the same time, p-MOSFET is turned off and some of the NBTI damage relaxes. Due to the fact that each degradation mechanism generates defects either in the bulk oxide or at the interface, the overall MOSFET degradation can be very complex.

8.1 NBTI in Strain-Engineered p-MOSFETs

State-of-the-art high-performance Si CMOS technologies rely on strain engineering, based on either a global approach using high-mobility substrates or implementation of local stressors [5]. Semiconductor manufacturers have successfully adopted strain engineering in 45 nm technology [6]. Local strain techniques are being adopted due to their low cost and ease in integration. Local stress may be induced by shallow trench isolation (STI), strained SiN cap layers, silicidation, and SiGe or SiC pockets. A compressive strain is introduced in the p-MOSFET channel by using embedded SiGe (e-SiGe) pockets in the source and drain region. Strain in the channel region affects device parameters such as negative bias temperature instabilities, low-frequency noise, radiation hardness, gate oxide quality, and hot-carrier performance. Recent reports [7] indicate that strain-engineered MOSFETs are prone to higher NBTI. In strained Si MOSFETs, due to crystal lattice mismatch at the Si/SiO_2 interface, traps are present in the form of a Si dangling bond. During fabrication, MOSFETs are annealed in hydrogen ambient to passivate the dangling Si bonds. Also, strain present at the Si/SiO_2 interface degrades the reliability by weakening the H_2-passivated Si dangling (Si-H) bonds by creating favourable conditions for interface state generation [8, 9]. The traps increase the threshold voltage, reduce the channel mobility due to higher scattering, induce parasitic capacitances in transistors, and lead to drain current degradation in the course of time.

In this section, we use Synopsys technology computer-aided design (TCAD) tools for simulation of trap generation at the Si/SiO_2 interface of p-MOSFETs with e-SiGe (SiGe pocket in the source and drain region) under negative gate bias. We account for the passivity of silicon dangling bonds by free hydrogen and its diffusion in the bulk oxide region and the activation energy of the Si-H bond process, which depends on the hydrogen concentration. The trap concentrations as a function of time in the bulk Si and process-induced strained Si p-MOSFET are compared. An analytical trap-induced Coulomb mobility model is developed and implemented in the SDevice simulator. We discuss in detail the influence of NBTI on the DC characteristics of strain-engineered p-MOSFETs.

8.1.1 Quasi-2D Coulomb Mobility Model

The strained Si/SiO_2 interface in strain-engineered p-MOSFETs shows a very large number of trap states [9]. These traps become filled during inversion, causing a change of conduction charge in the inversion layer and an increase in the Coulomb scattering of mobile charges. Owing to the large number of occupied interface traps, Coulomb interaction is likely to be an important scattering mechanism in process-induced strained Si p-MOSFET operation. Coulomb interaction results in very low surface mobilities and may be described by a quasi-2D scattering model. The Coulomb potential due to the occupied traps and fixed charges decreases with distance away from the interface. Mobile charges in the inversion layer that are close to the interface are scattered more than those farther away from the interface; therefore, the Coulomb scattering mobility model is required to be depth dependent. We assume that the electron gas can move in the x-y plane and is confined in the z direction. Electrons are considered confined or quantised if their deBroglie wavelength is larger than or comparable to the width of the confining potential. The deBroglie wavelength of electrons, given by $\lambda = \hbar/\sqrt{2m^*k_BT}$, is approximately 150 Å at room temperature, whereas the thickness of the inversion layer is typically around 50 to 100 Å. Thus, one may justify treating the inversion layer as a 2D electron gas.

The scattering from charged centres in the electric quantum limit has been formulated by Stern and Howard [10]. We consider only the p-channel inversion layer on the Si (100) surface where the Fermi line is isotropic and calculate the potential of a charged centre located at (r_i, z_i). Using the image method, we get

$$V_i(r, z) = \frac{e^2}{4\pi\varepsilon_0 \tilde{k}\sqrt{(r - r_i)^2 + (z - z_i)^2}} \tag{8.1}$$

where $r^2 = x^2 + y^2, z = 0$ corresponds the Si/SiO$_2$ interface. $z > 0$ is in silicon, whereas $z < 0$ is in the oxide, where $\tilde{k} = (k_{Si} + k_{ox})/2$ for $z < 0$, and ε_o is the permittivity of free space. We assume parabolic subbands with the same effective heavy-hole mass, m^*. Since inversion layer holes are restricted to move in the x-y plane, they would only scatter off potential perturbations that they see in the x-y plane. Therefore, we are only interested in determining the potential variations along that plane. To do so, one needs to calculate the 2D Fourier transforms of the potential appearing in Equation (8.1). The hole wave functions are then given by

$$\psi_{i,k}(r, z) = \frac{1}{\sqrt{A}} \xi(z) e^{ik \cdot r} \tag{8.2}$$

where i represents the subband index and $k = (k_x, k_y)$ is the 2D wave vector parallel to the interface. $\xi(z)$ is the quantised wave function in the direction perpendicular to the interface, E_i is its corresponding energy, and $r = (x, y)$. We denote the area of the interface by A. The effective unscreened quantum potential for holes in the inversion layer in the electric quantum limit in terms of the 2D Fourier transform is given by

$$v(q, z_i) = \frac{e^2}{2\tilde{k}\varepsilon_0 q} \iint \xi_i(z)\xi_j(z) e^{-q \cdot |z - z_i|} \, dz \tag{8.3}$$

We now consider the effect of screening due to inversion layer electrons on Coulombic scattering. Screening is actually a many-body phenomenon since it involves the collective motion of the electron gas. Using the Coulomb screening we get

$$v(q, z_i) = \frac{e^2}{2\tilde{k}\varepsilon_0 (q + q_s)} \iint \xi_i(z)\xi_j(z) e^{-q \cdot |z - z_i|} \, dz \tag{8.4}$$

where $q_s = \frac{e^2}{2\tilde{k}\varepsilon_0} \iint \xi_i(z)\xi_j(z) e^{-q \cdot |z - z_i|} dz$. . One can obtain the scattering rate using Fermi's golden rules,

$$S(q, z_i) = \frac{2\pi}{\hbar^2} \left(\frac{e^2}{2\tilde{k}\varepsilon_0 (q + q_s)} \iint \xi_i(z)\xi_j(z) e^{-q \cdot |z - z_i|} \, dz \right)^2 \delta(E_k - E_{k'}) \tag{8.5}$$

where $\hbar$ is Planck's constant. E_k and $E_{k'}$ denote the initial and final energies of the mobile charge being scattered. Scattering of inversion layer mobile charges takes place due to Coulombic interactions with occupied traps at the interface and also with fixed charges distributed in the oxide. We define the 2D charge density $N_{2D}\delta(z_i)$ at depth z_i inside the oxide as the combination of the fixed charge N_f and trapped charge N_{it} as

$$N_{2D}(z_i) = \begin{cases} N_{it} + N_f(0), & z_i = 0 \\ N_f(z_i), & z_i < 0 \end{cases} \tag{8.6}$$

Using the above approximation, one obtains the total transition rate. Since Coulombic scattering is an elastic scattering mechanism, the scattering rate or, equivalently, the inverse of the momentum relaxation time is then calculated as

$$\frac{1}{\tau_m} = \frac{N_{2D}(z_i)}{(2\pi)^2} \cdot \frac{2\pi}{\hbar^2} \left(\frac{e^2}{2\tilde{k}\varepsilon_0} \right)^2 \int \left(\frac{1}{(q + q_s)} \iint \xi_i(z)\xi_j(z) e^{-q|z-z_i|} dz \right)^2$$

$$\delta(E_k - E_{k'})(1 - \cos\theta)\delta k \tag{8.7}$$

Using the above relaxation time, one obtains the mobility of the ith subband as

$$\mu_i = \frac{e}{m^*} \cdot \frac{\displaystyle\int \sum_i \tau_m \varepsilon \frac{\partial f_0(\varepsilon)}{\partial \varepsilon} d\varepsilon}{\displaystyle\int \varepsilon \frac{\partial f_0(\varepsilon)}{\partial \varepsilon} d\varepsilon} \tag{8.8}$$

The average mobility, $\bar{\mu}$, is then given by [11]

$$\bar{\mu} = \frac{\displaystyle\sum_i p_i \mu_i^2}{\displaystyle\sum_i p_i \mu_i} \tag{8.9}$$

where p_i is the hole concentration in the ith subband. Taking into account the different scattering mechanism and using Matthiessen's rule, one obtains the total mobility μ. In the presence of the NBTI effect $N_{2D}(z_i)$ changes to $\Delta N_{2D}(z_i, t)$. This NBTI-induced change of interface traps degrades the mobility of the carriers in the channel of the MOS device and leads to a reduction in channel conductance and transconductance. The mobility model described above has been implemented in the SDevice simulator. To activate the mobility model, appropriate mobility values were defined in the fields of the parameter file. Mainly, the NBTI effect is based on interface trap generation due to broken Si-H bonds at the semiconductor/oxide interface, which is a rough surface where the highly ordered crystalline strained Si channel and the amorphous SiO_2 dielectric meet. The interface traps generated during NBTI lead to parametric shifts at the MOSFET level. These interface states can shift the threshold voltage of the strain-engineered p-MOSFET as

$$\Delta V_T(t) = \frac{t_{ox}}{\varepsilon_{ox}} \Delta N_{2D}(z_i, t) \qquad (8.10)$$

where ε_{ox} is the dielectric constant of the SiO_2 and t_{ox} is the thickness of the oxide.

8.2 Simulation of NBTI in p-MOSFETs

The simulations are performed on P⁺-poly-gate p-MOSFETs. NBTI stress is applied to the devices with negative gate bias at different temperatures. The temperatures are varied from 300 to 400°C. The interface trap generation for bulk Si p-MOSFETs and PSS p-MOSFETs at two different temperatures is shown in Figure 8.1.

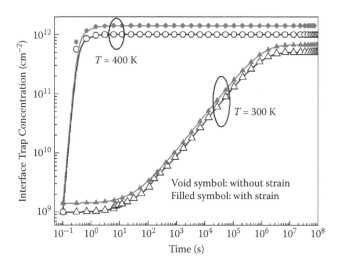

FIGURE 8.1
Temperature dependence of interface state degradation with the change in drain stress bias.

N_{it} generation depends on the curvature (or the interfacial strain at the Si substrate) and suggests the existence of strain (compressive strain) at the Si/SiO$_2$ interface. Strained bonds and bond defects such as Si dangling bond or Si-H bond in the network of SiO$_2$ films are responsible for the generation of hole traps that are mostly distributed near the Si/SiO$_2$ interface [10]. The compressive strain reduces the interatomic distance on the SiO$_2$ side more than that on the Si substrate side at the Si/SiO$_2$ interface region when the reference wafer turns from convex to concave. Therefore, the bond mismatch at the Si/SiO$_2$ interface becomes very small. Thus, N_{it} generation is less due to the marked variation of bond mismatch at the Si/SiO$_2$ interface. N_{it} generation is also reaction-diffusion (R-D) limited.

The Si dangling bond generated can consequently be traced to be at a new equilibrium position with a different bond reformation probability. Since a drastic change of interatomic distance does not occur within the Si/SiO$_2$ interface due to compressive strain, the bond reformation efficiency is high and broken bonds are repaired in the region of compressive strain [12]. Thus, the results indicate that compressive strain does not create favourable conditions for additional interface state creation. Figure 8.1 shows that interface state generation for drain bias V_{ds} = 50 mV, which depends on temperature, with N_{it} being higher for devices stressed at a higher temperature.

A change in the initial I_{ds}-V_{gs} dependency, induced by the traps accumulated in the channel, may be seen in Figure 8.2. A higher temperature results in a larger trap concentration (characteristic of NBTI), and for a particular temperature, computation accounting for the diffusion of the free hydrogen into the oxide layer also results in a larger trap concentration, and

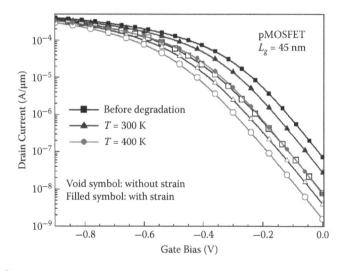

FIGURE 8.2
Drain current (*Ids*) vs. gate voltage (*V$_{gs}$*) for drain bias V_{ds} = 50 mV, before and after the degradation simulation for bulk Si and process-induced strained Si p-MOSFETs.

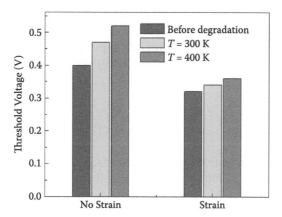

FIGURE 8.3
Change in threshold voltage during NBTI for p-MOSFETs simulated at different temperature.

hence more degradation in drain current. Figure 8.3 shows the behaviour of the threshold voltage change. Total degradation can be modeled with an effective threshold voltage shift since both mobility and ΔV_T are proportional to $\Delta N_{2D}(z_i)$.

8.3 HCI in Strain-Engineered n-MOSFETs

As the device dimensions are scaled, short-channel effects become more important. Hot carrier is an important short-channel phenomenon resulting from the high electric field in the device and causes degradation in device characteristics [13]. Thus, studies on the degradation mechanisms of DC parameters have received serious attention. It is known that the degradation of MOS transistors is caused by interface trap generation resulting from hot-carrier injection. Contrary to the traditionally accepted concept that interface traps are generated mostly by the hot hole and electron injection into the oxide, it has been shown recently that during hot-carrier stressing, channel hot electrons, which are not injected into the oxide, generate a significant amount of interface damage [14]. In this section, DC hot-carrier-induced degradation is discussed based on the studies using simulation. Trap formation kinetics in the channel of n-MOSFETs has been studied, and it has been shown that the traps are generated due to the breaking of Si-H bonds.

8.3.1 Degradation Mechanisms

Device reliability is a serious concern and includes various degradation mechanisms: trapping of injected charge, trap generation, oxide breakdown, hot-carrier effects, ion drift, and interdiffusion of metals, stress migration, and mechanical effects [15]. It is well known that H plays a critical role in the fabrication of high-quality Si/SiO$_2$ interfaces where these dangling bonds are compensated by hydrogen atoms. The experimental data for the kinetics of interface trap formation show that the time dependence of trap generation can be described by the relation

$$N_{it} - N_{it}^0 = \frac{N_{hb}^0}{1+(\gamma t)^\alpha}$$ (8.11)

where N_{it} is the concentration of the interface, and $N^0{}_{hb}$ and $N^0{}_{it}$ are the initial concentrations of the Si-H bonds and interface traps, respectively. Considering $N = N_{hb}^0 + N_{it}^0$ total Si bonds at the interface, the remaining number of Si-H bonds at the interface after stress is $N_{hb} = N - N_{it}$, given by

$$N_{hb} = \frac{N_{hb}^0}{1+(\gamma t)^\alpha}$$ (8.12)

and the Si-H concentration during stress is given by

$$\frac{dN_{hb}}{dt} = -\gamma . N_{hb}$$ (8.13)

where γ is a reaction constant and is given by $\gamma = \gamma_0 \exp(-\varepsilon_A/k_B T)$ in the Arrhenius approximation. ε_A is the Si-H activation energy and T is the temperature. The activation energy needed to release hydrogen from the interface can be expressed as

$$\varepsilon_A = \varepsilon_A^0 + (1+\beta)kT \ln\left(\frac{N - N_{hb}}{N - N_{hb}^0}\right)$$ (8.14)

where ε_A is the energy needed to break a Si-H bond and the last term represents the potential energy needed to go over the potential barrier of the 2D potential system with prefactor β.

8.4 Simulation of HCI in n-MOSFETs

In the following, simulation results for the strain-engineered n-MOSFETs after hot-carrier stressing are presented. The constant voltage stress (electrical stress) conditions for which the simulations are performed are gate voltage (V_{gs}) range of 1.5 to 2.8 V, and the device is kept under the stress for 10^5 s. When the degradation simulation is finished, the device is set to normal

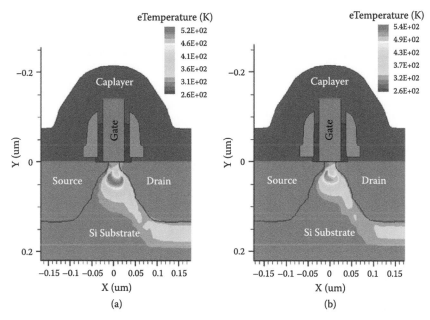

FIGURE 8.4
Electron temperature distribution in strain-engineered n-MOSFETs after hot-carrier stressing.

operating conditions and the I_{ds}-V_{gs} sweep is performed again. Simulated electron temperature distribution due to the electrical stressing is presented in Figure 8.4 for both the electric field stressing and hot-carrier stressing for devices with (a) a relax cap layer and (b) a highly tensile cap layer.

The trap generation profiles along the Si/SiO$_2$ interface are shown in Figure 8.5. As expected, electrically stressed devices show higher interface traps.

Reliability comparison for strained devices was systematically studied in simulation by applying different gate and drain voltages. Important simulation results are presented below. The simulated drain current as a function of the gate voltage, transconductance vs. gate voltage, and drain current as a function of drain voltage before and after stressing are shown in Figures 8.6 to 8.8, respectively.

The threshold voltage extracted using maximum transconductance method, before and after degradation, is shown in Table 8.1.

TABLE 8.1

Comparison of Threshold Voltage

Type	Threshold Voltage before Degradation (V)	Threshold Voltage after Degradation (V)
Without strain	0.331	0.347
With strain	0.305	0.328

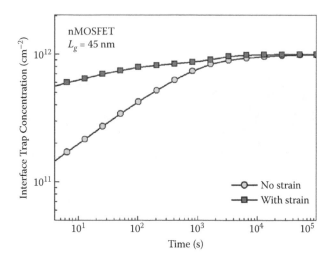

FIGURE 8.5
Trap distribution at the interface.

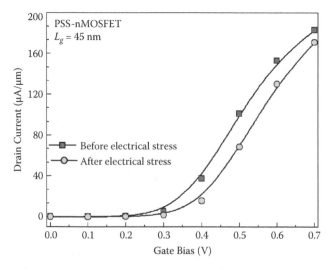

FIGURE 8.6
Simulated gate voltage vs. drain current (I_{ds}-V_{gs}) characteristics for process-induced strained Si n-MOSFETs showing device degradation. A decrease in drain current is observed.

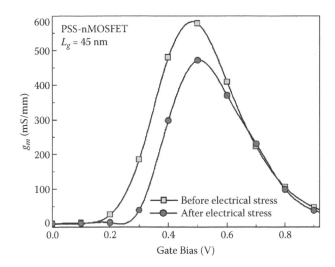

FIGURE 8.7
Simulated transconductance showing device degradation. Decrease in transconductance is observed.

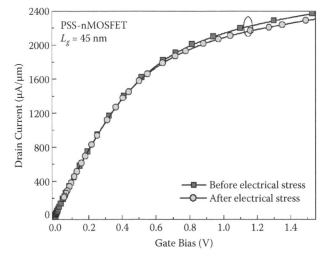

FIGURE 8.8
Simulated I_{ds}-V_{ds} characteristics.

8.5 Reliability Issues in FinFETs

Most integrated circuits are fabricated on (100)-oriented Si wafers. However, it has been known since 1968 that hole mobility is higher for p-MOSFETs on (110)-oriented wafers with the channel in the <110> direction. Along with the higher hole mobility, however, the (110) Si surface has a higher Si bond availability. This increases the probability of de-passivated Si bonds, and one would expect more severe NBTI degradation, as indeed has been observed [16]. This is a potential problem if (110)-oriented wafers become important. For certain 3D devices, e.g., FinFETs, when fabricated on (100) wafers with the channel in the conventional <110> direction, the vertical sidewalls are (110) oriented, leading to NBTI problems. However, forming FinFETs on (100) wafers with the channel in the <100> direction leads to (100) vertical sidewalls. In the case of the triple-gate devices, approximately two-thirds of the total active area is on the lateral (110) sides, whereas in the case of planar devices almost all the active area is on the top (100)-oriented side. (110)-oriented Si surfaces are known to have a higher density of available Si bonds, and thus a higher density of interface states. The authors reported that the threshold voltage shift caused by negative bias temperature stress for triple-gate transistors was worse than that in planar devices and attributed this effect to the larger trap density of the (110) sidewall channel. In order to study this effect, 45° rotated notch structures, having (100) crystal orientation on both top and sidewalls, are used for comparison. Shickova [17] has addressed the effects of additional processing steps required to introduce strain, showing that their effects need to be considered in order to make a valid NBTI comparison, ensuring that the compared devices are stressed at the same E_{ox}.

The new high-k dielectrics also contribute to the already increased threshold voltage shifts. Thus, a proper passivation of the dielectric is critical in order to overcome these problems. Passivation by fluorine as a possible means to reduce the number of interface and bulk defects is an attractive alternative to hydrogen passivation, and it was the subject of several recent studies. Another concern of the reliability of the multigate devices is the already mentioned "corner effect," caused by the concentration of the electric field around the fin corners. This local increase of the electric field may lead to preferential breakdown at the fin corners, in case an appropriate corner rounding processing has not been used [18, 19]. A systematic and comprehensive study of the impact of process-induced strain on NBTI reliability has been reported, including devices with different gate stacks, as well as different strain introduction techniques. The study included devices with different gate stacks as well as different strain introduction techniques. Gate stacks studied include poly-Si/SiON, TiN/HfO$_2$/SiO$_2$, and Ni fully silicided gates (FUSI)/HfSiON/SiO$_2$. Strain introduction techniques include compressive stressor layers (contact etch stop layers (CESLs)) and SiGe S/D. Two

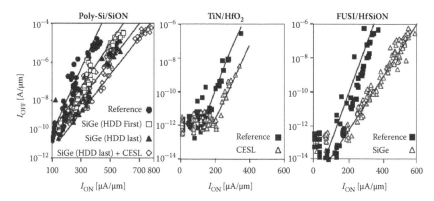

FIGURE 8.9
I_{on}-I_{off} plots, showing significant improvement in drive current for p-MOSFET devices with process-induced compressive strain for different gate stacks. (After Shickova, A., Bias Temperature Instability Effects in Devices with Fully-Silicided Gate Stacks, Strained-Si, and Multiple-Gate Architectures, PhD thesis, Katholieke Universiteit Leuven, 2008.)

different process sequences were studied for SiGe S/D, changing the step sequence of junction formation and selectively epitaxial growth (SEG).

Interrupted NBTI measurements were complemented in this study with charge pumping and noise measurements in order to obtain a more complete view of defects present and their generation under stress. The effects of the processing steps on electrical properties of the devices were carefully taken into account, allowing us to make a fair comparison between devices with and without strain. NBTI characterisation results are presented below. I_{on}-I_{off} characteristics for the different devices are shown in Figure 8.9. Significant performance improvement is seen for devices with compressive process-induced strain, confirming the presence and impact of the strain.

Figure 8.10(a) shows the threshold voltage dependence on stress time for a fixed gate voltage in the case of devices with poly-Si/SiON gate stacks. An apparent improvement in NBTI is observed for the SiGe (Highly Doped Drain [HDD] first) case, while a slight apparent deterioration is observed in the case of SiGe (HDD last) + CESL. After correcting the gate stress voltages to produce the same oxide electric fields, no significant difference is observed anymore in the NBTI time dependence plots (Figure 8.10(b)). Figure 8.11 shows the threshold voltage dependence on stress time at fixed oxide electric fields in the cases of the FUSI/HfSiON and TiN/HfO$_2$ gate stacks. As in the case of poly-Si/SiON stacks, when compared to reference devices at the same electric field, devices with process-induced strain show identical behaviour and no strain-induced degradation is observed. Figure 8.12 shows the results of the lifetime extrapolations. No degradation was observed due to the presence of strain in any of the gate stacks reported.

In the case of a FinFET, preferential breakdown at fin edges has been a concern for the 3D architecture. The study on time-dependent dielectric

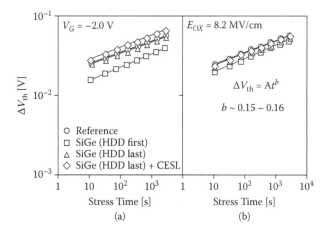

FIGURE 8.10
Threshold voltage shifts vs. stress time $\Delta V_{th}(t)$: (a) plotted in the conventional way at a fixed stress gate voltage of $VG = -2.0$ V, showing apparent improvement of NBTI for the SiGe (HDD first) case and a slight deterioration in the SiGe (HDD last) + CESL case and (b) plotted at a fixed oxide electric field $E_{ox} = 8.2$ MV/cm, showing similar NBTI results between the devices with strain and reference devices. (After Shickova, A., Bias Temperature Instability Effects in Devices with Fully-Silicided Gate Stacks, Strained-Si, and Multiple-Gate Architectures, PhD thesis, Katholieke Universiteit Leuven, 2008.)

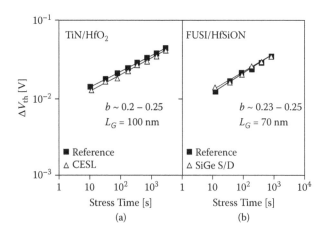

FIGURE 8.11
Threshold voltage shifts vs. stress time $\Delta V_{th}(t)$ plotted at a fixed oxide electric field: (a) TiN/HfO$_2$ at $E_{ox} = 7.3$ MV/cm and (b) FUSI/HfSiON at $E_{ox} = 8$ MV/cm. The NBTI results were similar between the devices with strain and reference devices. (After Shickova, A., Bias Temperature Instability Effects in Devices with Fully-Silicided Gate Stacks, Strained-Si, and Multiple-Gate Architectures, PhD thesis, Katholieke Universiteit Leuven, 2008.)

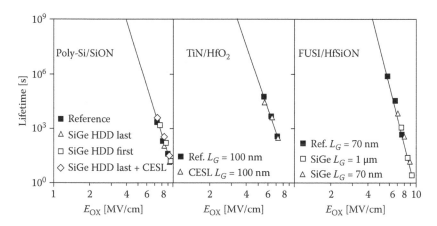

FIGURE 8.12
Lifetime extrapolations for devices with process-induced strain and reference devices for a degradation criterion of 30 mV V_{th} shift, assuming a power law dependence on the oxide electric field. $T = 125°C$. (After Shickova, A., Bias Temperature Instability Effects in Devices with Fully-Silicided Gate Stacks, Strained-Si, and Multiple-Gate Architectures, PhD thesis, Katholieke Universiteit Leuven, 2008.)

breakdown showed a significantly degraded Weibull shape factor, β, and voltage acceleration factor, γ, for the multifin devices without corner rounding. When adequate corner rounding is applied, however, multigate devices show breakdown behaviour similar to that of planar devices. In order to decouple the effects of the Si crystal orientation from the effects of nitridation, 45° rotated structures resulting in (100) Si at both top and sidewalls were studied (Figure 8.13).

Figure 8.14 shows BTI comparisons between reference (nonnitrided) devices and devices with ammonia nitridation for 45° rotated notch wafers,

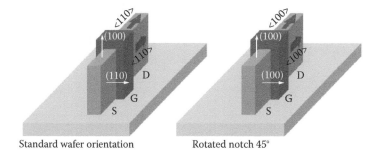

FIGURE 8.13
Top and sidewall Si crystal orientations (and current directions) in the cases of standard wafer orientation and 45° rotated notch. (After Shickova, A., Bias Temperature Instability Effects in Devices with Fully-Silicided Gate Stacks, Strained-Si, and Multiple-Gate Architectures, PhD thesis, Katholieke Universiteit Leuven, 2008.)

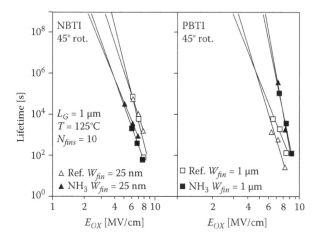

FIGURE 8.14

BTI comparisons between reference (nonnitrided) devices and devices with ammonia nitrid-ation for 45° rotated notch wafers, for both narrow and wide-fin devices. Very similar BTI behaviour is observed for narrow and wide-fin devices. Improved PBTI and degraded NBTI are consistent with the data obtained on planar devices. (After Shickova, A., Bias Temperature Instability Effects in Devices with Fully-Silicided Gate Stacks, Strained-Si, and Multiple-Gate Architectures, PhD thesis, Katholieke Universiteit Leuven, 2008.)

for both narrow- and wide-fin devices. Very similar BTI behaviour is observed for wide- and narrow-fin devices (for both reference devices and nitrided devices). The improved PBTI and degraded NBTI behaviour with nitridation observed in Figure 8.14 is consistent with the data obtained on planar devices and indicates nitrogen incorporation into the bulk of the dielectric, as well as at the Si interface. Since the behaviour of narrow-fin devices is dominated by the behaviour of the fin sides, and that of wide-fin devices by the top, the negligible differences in BTI between narrow- and wide-fin devices with nitridation indicate a similar distribution of the nitro-gen in the dielectric at the sides and top of the fins, both in the bulk and close to the Si/SiO_2 interfaces.

8.6 Summary

Negative bias temperature instability in p-MOSFETs and hot-carrier injec-tion in n-MOSFETs are serious reliability concerns for digital and analogue CMOS circuit applications. In this chapter, effects of strain in the channel region on negative bias temperature instabilities, gate oxide quality, and hot-carrier performance have been discussed in detail from fundamental physics. Technology CAD has been used to study the effects of strain on the negative bias temperature instabilities in p-MOSFETs and hot-carrier injection in n-MOSFETs.

Review Questions

1. What is NBTI?
2. What is HCI?
3. What is bias temperature instability?
4. NBTI is characterised by an increase in threshold voltage. (True/False)
5. The NBTI mechanism is due to breaking of Si-H bonds at the SiO_2/Si interface. (True/False)
6. Negligible PBTI effects are observed in n-MOSFETs. (True/False)
7. High-k dielectrics contribute to threshold voltage shift. (True/False)
8. What are the effects of strain on NBTI?
9. Interfacial defects intrinsic to the presence of strain affect BTI. (True/False)
10. An increase in defects increases leakage current through the gate dielectric. (True/False)

References

1. Y. Miura and Y. Matukura, Investigation of Silicon–Silicon Dioxide Interface Using MOS Structure, *Jpn. J. Appl. Phys.*, 5, 180–180, 1966.
2. M. A. Alam and S. Mahapatra, A Comprehensive Model of PMOS NBTI Degradation, *Microelectronics Reliability*, 45, 71–81, 2005.
3. S. Mahapatra, D. Saha, D. Varghese, and P. B. Kumar, On the Generation and Recovery of Interface Traps in MOSFETs Subjected to NBTI, FN and HCI Stress, *IEEE Trans. Electron Dev.*, 53, 1583–1591, 2006.
4. H. Kufluoglu and M. A. Alam, A Geometrical Unification of the Theories of NBTI and HCI Time-Exponents and Its Implications for Ultra-Scaled Planar and Surround Gate MOSFETs, *IEEE IEDM Tech. Dig.*, 113–116, 2004.
5. C. K. Maiti, S. Chattopadhyay, and L. K. Bera, *Strained-Si Heterostructure Field Effect Devices*, CRC Press, Boca Raton, FL, 2007.
6. S. Thompson, M. Armstrong, C. Auth, M. Alavi, M. Buehler, R. Chau, S. Cea, T. Ghani, G. Glass, T. Hoffman, C.-H. Jan, C. Kenyon, J. Klaus, K. Kuhn, Zhiyong Ma, B. Mcintyre, K. Mistry, A. Murthy, B. Obradovic, R. Nagisetty, P. Nguyen, S.

Sivakumar, R. Shaheed, L. Shifren, B. Tufts, S. Tyagi, M. Bohr, and Y. El-Mansy. A 90-nm Logic Technology Featuring Strained-Silicon, *IEEE Electron Dev. Lett.*, 51, 1790–1797, 2004.

7. C.-H. Liu and T.-M. Pan, Hot Carrier and Negative-Bias Temperature Instability Reliabilities of Strained-Si MOSFETs, *IEEE Trans. Electron Dev.*, 54, 1799–1803, 2007.

8. A. Shickova, B. Kaczer, P. Verheyen, G. Eneman, E. San Andres, M. Jurczak, P. Absil, H. Maes, and G. Groeseneken, Negligible Effect of Process-Induced Strain on Intrinsic NBTI Behavior, *IEEE Electron Dev Lett.*, 28, 242–244, 2007.

9. G. K. Dalapati, S. Chattopadhyay, K. S. K. Kwa, S. H. Olsen, Y. L. Tsang, R. Agaiby, A. G. O'Neill, P. Dobrosz, and S. J. Bull, Impact of Strained-Si Thickness and Ge Out-Diffusion on Gate Oxide Quality for Strained-Si Surface Channel n-MOSFETs, *IEEE Trans. Electron Dev.*, 53, 1142–1152, 2006.

10. F. Stern and W. E. Howard, Properties of Semiconductor Surface Inversion Layers in the Electric Quantum Limit, *Phys. Rev.*, 163, 816–835, 1967.

11. K. Inoue and T. Matsuno, Electron Mobilities in Modulation-Doped AlxGa1-xAs/GaAs and Pseudomorphic AlxGa1-xAs/InyGa1-yAs Quantum-Well Structures, *Phys. Rev. B Condens. Matter*, 47, 3771–3778, 1993.

12. T. K. Maiti, S. S. Mahato, S. K. Sarkar, and C. K. Maiti, Impact of Negative Bias Temperature Instability on Strain-Engineered p-MOSFETs, *Proceedings of International Conference on Materials for Advanced Technologies (ICMAT 2007)*, Singapore, E-13-OR52, 2007.

13. K.-W. Ang, C. Wan, K.-J. Chui, C.-H. Tung, N. Balasubramanian, M.-F. Li, G. Samudra, and Y.-C. Yeo, Hot Carrier Reliability of Strained N-MOSFET with Lattice Mismatched Source-Drain Stressors, *Proc. IRPS*, 684–685, 2007.

14. T. K. Maiti, S. S. Mahato, M. K. Bera, M. Sengupta, P. Chakraborty, C. Mahata, A. Chakraborty, and C. K. Maiti, Stress-Induced Degradation in Strain-Engineered nMOSFETs, *Proc. Int. Symp. Physical Failure Anal. Integrated Circuits (IPFA 2008)*, 1–3, 2008.

15. D. Lachenal, F. Monsieur, Y. Rey-Tauriac, and A. Bravaix, HCI Degradation Model Based on the Diffusion Equation Including the MVHR Model, *Microelectronic Eng.*, 84, 1921–1924, 2007.

16. S. Maeda, J.-A. Choi, J.-H. Yang, Y.-S. Jin, S.-K. Bae, Y.-W. Kim, and K.-P. Suh, Negative Bias Temperature Instability in Triple Gate Transistors, *Proc. IRPS*, 8–12, 2004.

17. A. Shickova, Bias Temperature Instability Effects in Devices with Fully-Silicided Gate Stacks, Strained-Si and Multiple-Gate Architectures, PhD thesis, Katholieke Universiteit Leuven, 2008.

18. A. Shickova, N. Collaert, R. Rooyackers, A. De Keersgieter, T. Kauerauf, M. Jurczak, B. Kaczer, and G. Groeseneken, Dielectric Breakdown Study of Multi-Gate Devices, *Proc. ULIS*, 141–144, 2006.

19. A. Shickova, P. Verheyen, G. Eneman, R. Degraeve, E. Simoen, P. Favia, D. O. Klenov, E. S. Andrés, B. Kaczer, M. Jurczak, P. Absil, H. E. Maes, and G. Groeseneken, Reliability of Strained-Si Devices with Post-Oxide-Deposition Strain Introduction, *IEEE Trans. Electron Dev.*, 55, 3432–3441, 2008.

9

Process Compact Modelling of Strain-Engineered MOSFETs

As complementary metal-oxide-semiconductor (CMOS) downscaling approaches the manufacturing limits, process variability and reliability degradation become the key limiting factors for future integrated circuits and
systems design. At nanoscale, physical factors that previously had little or no
impact on circuit performance are now becoming increasingly significant.
Examples include process variations, transistor mobility degradation, and
power consumption. These new effects pose dramatic challenges to robust circuit design and system integration. Process variations have become increasingly important for scaled technologies starting at 45 nm, as nontraditional
materials and structures and even strain technology are being introduced to
enhance the device performance. Use of strain technology in manufacturing has urged that the designers assess layout-dependent effects and man-
age their impact. Thus, the demand of predictive modelling becomes even
stronger as we face more complicated and diverse technological choices for
larger-scale integration. High process variability not only affects the circuit
performance but also reduces manufacturing yield. To improve manufacturing yield of technologies 45 nm and below, performance variability should
be considered during the design phase. In the conventional design approach,
high variability leads to overdesigning, thereby increasing area and power
consumption. To avoid overdesigning, accurate estimation of variability is
required.

According to the International Technology Roadmap for Semiconductors
(ITRS), for technology nodes beyond 45 nm, larger amounts of process variations are expected. The increased variations are primarily due to random
dopant fluctuations, line-edge roughness, and oxide thickness fluctuation.
These variations greatly impact all aspects of circuit performance and pose
a great challenge to future integrated circuit (IC) design. To improve robustness, efficient methodology is required that considers the effect of variations
in the design flow. What matters is not only the amount of variations, but
also the sensitivity to variations. At the nanoscale, the sensitivity of transistor performance on process variations becomes more significant and is critical for robust CMOS design.

This chapter covers both the modelling principles and the applications
of predictive technology modelling (PTM) and process compact modelling (PCM) in microelectronics design. We discuss the methodology for

constructing compact SPICE models as a function of process parameter variations. We present a simulation methodology for strain-engineered metal-oxide-semiconductor field-effect transistors (MOSFETs), which allow the flow of pertinent information between process and design engineers without the need for disclosing the details of process technology. The methodology involves global extraction of process-dependent SPICE model parameters. Linking design and process, statistical compact models provide the essential correlation between performance statistics and process parameter statistics.

9.1 Process Variation

Process variations refer to those variations caused due to the imperfections in different steps of the manufacturing process; these could be due to the limited resolution of the photolithographic stage within the fabrication process, which results in variations in the width and length of transistors on the chip. It could also be from nonuniform conditions during the diffusion stage, in which impurities are introduced. These imperfections cause variations in the electrical properties of the transistors and interconnect on the chip from their designed values. Examples are variations in the geometries of the transistors (e.g., effective channel length, oxide thickness), or due to random dopant fluctuations (affecting the threshold voltage of the transistors).

In general, the process variations can be distinguished into the following components:

1. Die-to-die (interdie) variation: These are largely independent of design implementation and cause systematic variations in electrical characteristics within the chip.
2. Within-die (intradie) variations. These can be distinguished into four subcategories:
   a. Wafer-level variations due to nonuniformities (e.g., thermal gradient)
   b. Die-level variations caused by imperfections in mask making/ lithography
   c. Wafer-die interaction on account of dependence due to chip location within the wafer
   d. Random residuals due to random dopant fluctuation, etc.

The first three are correlated systematic components, whereas the last one forms the uncorrelated random component in intradie process variation. Intradie variations are more difficult to solve because these variations are not systematic. Interdie variations are systematic and affect adjacent transistors on a chip with equal shift from nominal value. Intradie variations are random variations and affect adjacent transistors on the same chip with different shifts.

9.2 Predictive Technology Modelling

To continue the design success with nanoscale CMOS, one requires an early comprehension of the technology impacts on circuit design. Although high-k/metal gate and strained silicon techniques have helped extend the CMOS technology, they have also complicated the fabrication process and increased the amount of process variations. Aggressive technology scaling has led to large uncertainties in device and interconnect characteristics for deep-submicron circuits. Many physical phenomena, unforeseen in the larger dimensions, such as short-channel effect (SCE) and exponential increase in leakage, are becoming the major bottlenecks for continuous technology scaling. Increasing variations (both interdie and intradie) in device parameters (channel length, gate width, oxide thickness, device threshold voltage, etc.) produce a large spread in the delay and power consumption in advanced integrated circuits. The presence of large process variations and deep-submicron effects requires a paradigm change in the design and optimisation of large-scale circuits and systems. It is important to link the process parameters, including the distribution to SPICE parameters, to study the global variations at the circuit level. By using accurate physical models of the manufacturing process, custom designers can account for manufacturing variability.

For circuit design, it is critical to have predictive MOSFET models that are reasonably accurate, scalable, and correctly capture the new physical effects arising out of the nanoscale CMOS technology. Examples of the emerging challenges include leakage current, process variations, and transistor reliability. The predictive technology model (PTM) is critical for early circuit design research to assess performance trends and evaluate key modules to facilitate the development of future CMOS technology. It is currently being used to predict the characteristics of nanoscale CMOS, including physical effects, process variations, and physical correlations among model parameters. A new generation of predictive technology models for front-end-of-the-line (FEOL) CMOS technology has been developed from the 250 nm to 32 nm nodes, including both high-performance and low-power processes and alternative structures such as FinFET and high-k/metal gate (HK-MG) based on the Berkeley Predictive Technology Model (BPTM) [1].

Back-end-of-line (BEOL) interconnects become a limiting factor in circuit performance as complementary metal-oxide-semiconductor (CMOS) technology scales. Accurate and efficient modelling of back-end-of-line interconnect paratactic capacitance is essential in determining various on-chip interconnect-related issues such as delay, cross talk, resistive drop, and power dissipation. As CMOS technology continues scaling, metal wiring pitch concurs to shrink with transistor feature size to increase chip density. This makes BEOL metal wiring line capacitance (c) and resistance (R), i.e., the RC delay, difficult to be reduced fast enough, compared to the ever-increasing FEOL transistor speed. As a result, the interconnect parasitic becomes a limiting factor in circuit performance. Meanwhile, the interconnect structure becomes increasingly complex, and nonuniform dielectrics, such as stop layer and air gap, are being used. To efficiently extract the paratactic of interconnects, a compact capacitance model is developed.

The predictive technology model, which was initiated at the University of California, Berkeley, in 1999, bridges process development and circuit design through device modelling, and is essential for supporting early design prototyping. PTM is a critical interface between technology innovation and IC design exploration [2]. PTM introduces scalable models for strained Si, multiple V_{th}, and HK-MG processes, and even the FinFET structures. Primary parameters under the influence of these technology enhancements include the increase of mobility, the control of SCE, and the coupling between front and back gates in a FinFET device. PTM quantitatively evaluates various technology factors in scaled CMOS design, helping prediction on the performance trend along the road map. PTM has been used for a 45 nm predictive process design kit (PDK), which is the critical interface between circuit design and silicon fabrication.

Initially, PTM was proposed to help bridge the technology and design groups, such that these issues could be brought to attention as early as possible in the design process. The current Berkeley Predictive Technology Model (BPTM), based on the BSIM4 model, includes more physical parameters and provides a standard compact model down to the 12 nm technology node [1]. Recently PTM has been extended from conventional CMOS devices to advanced devices, including strained Si, HK-MG, and the double-gate structures. To predict future technology characteristics, however, a simple approach to scale down the geometry and voltages from an existing technology does not work. For example, a comparison of predicted device data based on a well-characterised 130 nm technology, when scaled down in terms of L_{eff}, T_{ox}, V_{th0}, R_{dsw}, and V_{dd} for an early 65 nm technology device, shows an overall performance underestimation (Figure 9.1).

Although there are typically more than 100 parameters in a compact transistor model to calculate the current-voltage (I-V) and capacitance-voltage (C-V) characteristics, only about 10 of them are critical to determine the essential behaviour of nanoscale transistors. The accuracy of PTM predictions has been verified with published silicon data; an error in I_{on} is below 10% for both n- and p-MOSFET devices. By tuning only 10 primary model parameters,

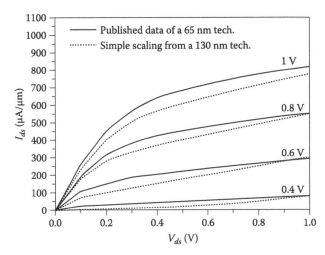

FIGURE 9.1
A simple method fails to predict the overall I-V characteristics. (After Zhao, W., Predictive Technology Modelling for Scaled CMOS, PhD thesis, Arizona State University, 2009.)

PTM can be easily customised to cover a wide range of process uncertainties. Furthermore, PTM captures the sensitivity to process variations.

Combining 10 primary parameters, e.g., V_{dd}, T_{oxe}, L_{eff}, V_{th0}, R_{dsw}, N_{ch}, E_{ta0}, K_1, m_0, and V_{sat}, PTM can be extrapolated toward future technology nodes. The rest of the model parameters of the compact transistor model are secondary ones. There are no explicit models to predict their values. Furthermore, their values can be adjusted to cover a range of process uncertainties from intrinsic process variations. In general, the error introduced by considering only the primary parameters can be reduced to 5% [3]. Smooth and accurate predictions are obtained from 250 nm to 32 nm nodes, with L_{eff} as low as 13 nm. The new predictive methodology reported has better scalability over a wide range of process and design conditions.

Based on the collected data, Figure 9.2 presents the trend of equivalent oxide thickness (EOT). EOT is steadily scaling down, although the pace may slow down. The trend of V_{dd} and V_{th} scaling is plotted in Figure 9.3, where the value of V_{th} is extracted from the subthreshold I-V curves. Due to the concern of subthreshold leakage, V_{th} almost stays the same in the nanoscale. The fifth technology parameter, R_{dsw}, is extracted by fitting the I-V curves in the linear region, after low-filed mobility, μ_o, is predicted. The trend of R_{dsw} is shown in Figure 9.4. The reduction of R_{dsw} becomes more difficult in short-channel devices. These trends, which are supported by experimental data, are then integrated into PTM to predict the nominal values during CMOS technology scaling.

Values of technology specifications not only define the basic characteristics of a process, but also further determine other important electrical

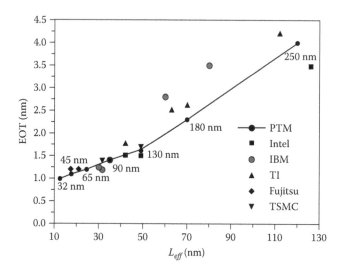

FIGURE 9.2
The trend of EOT scaling from the 250 nm to 32 nm nodes. (After Zhao, W., Predictive Technology Modelling for Scaled CMOS, PhD thesis, Arizona State University, 2009.)

details of a transistor. In particular, channel doping concentration, N_{ch}, is mainly defined by the threshold voltage. The exact value of N_{ch} is reversed from published data of V_{th0}, using the V_{th} model in BSIM [4]. Figure 9.5 illustrates the trend of N_{ch} scaling. Based on N_{ch}, the main coefficient for the body effect of V_{th}, K_1, is also estimated. Furthermore, to model the

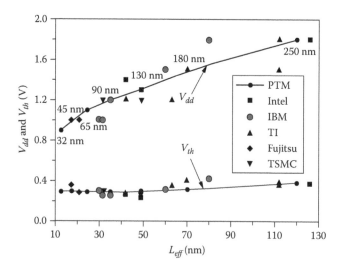

FIGURE 9.3
The trend of V_{dd} and V_{th} scaling from the 250 nm to 32 nm nodes. (After Zhao, W., Predictive Technology Modelling for Scaled CMOS, PhD thesis, Arizona State University, 2009.)

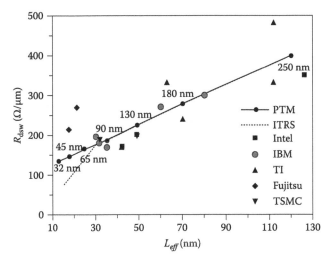

FIGURE 9.4
Trend of R_{dsw} scaling from the 250 nm to 32 nm nodes. (After Zhao, W., Predictive Technology Modelling for Scaled CMOS, PhD thesis, Arizona State University, 2009.)

V_{th} behaviour of short-channel transistors, drain-induced barrier lowering (DIBL) needs to be accounted for. To the first order, this effect is captured by E_{ta0}, which is a model parameter for the DIBL effect. Its value is extracted from published data of V_{th} roll-off, and the trend of E_{ta0} is illustrated in Figure 9.6.

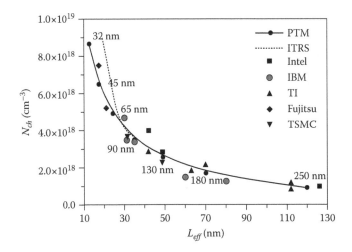

FIGURE 9.5
Trend of N_{ch} scaling from the 250 nm to 32 nm nodes. (After Zhao, W., Predictive Technology Modelling for Scaled CMOS, PhD thesis, Arizona State University, 2009.)

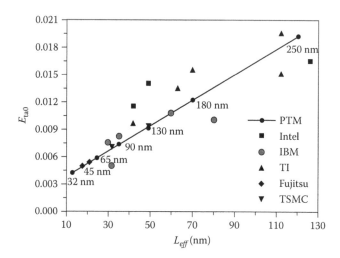

FIGURE 9.6
Trend of DIBL coefficient E_{ta0} scaling from the 250 nm to 32 nm nodes. (After Zhao, W., Predictive Technology Modelling for Scaled CMOS, PhD thesis, Arizona State University, 2009.)

Based on the successful verifications, PTMs for 130 nm to 32 nm technology generations have been generated. Figure 9.7 illustrates the trend of nominal I_{on} and I_{off}. Figure 9.8 illustrates the trend of nominal CV/I and switch power (CV_{aa}^2).

The overall map of process sensitivities is shown in Figure 9.9 across technology generations from 250 nm to 32 nm. Due to increasing process

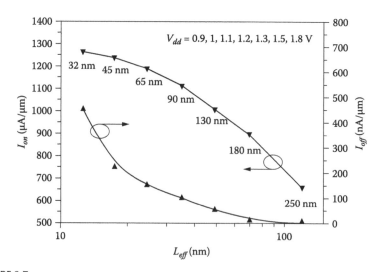

FIGURE 9.7
PTM nominal predictions of I_{on} and I_{off} from 250 nm to 32 nm nodes. (After Zhao, W., Predictive Technology Modelling for Scaled CMOS, PhD thesis, Arizona State University, 2009.)

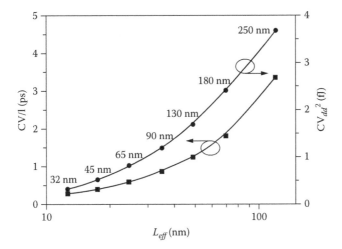

FIGURE 9.8
Prediction of nominal CV/I and CV_{dd}^2 from 250 nm to 32 nm nodes. (After Zhao, W., Predictive Technology Modelling for Scaled CMOS, PhD thesis, Arizona State University, 2009.)

sensitivities, the variation of I_{on} becomes larger during technology scaling, even if the normalised process variation remains constant, e.g., −20% and −12% for L_{eff} and N_{ch} variations, respectively (Figure 9.9). For future technology generations, L_{eff} will continue to be the dominant factor affecting performance variation, because of its role in velocity overshoot and the DIBL effect.

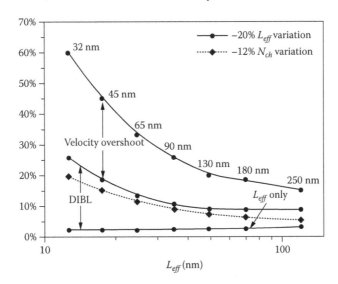

FIGURE 9.9
The impact of L_{eff} variation on I_{on} across technology generations. (After Zhao, W., Predictive Technology Modelling for Scaled CMOS, PhD thesis, Arizona State University, 2009.)

Second to L_{eff} variation, the impact of N_{ch} variation also keeps increasing as technology scales. Figure 9.9 shows the decomposition of the impact of L_{eff} variations during technology scaling. It reveals that velocity overshoot plays a more important role than DIBL for nanoscale MOSFET. Therefore, physical modelling of velocity overshoot is necessary in variation-aware design. Since PTM can be easily customised by tuning L_{eff}, T_{Oxe}, R_{dsw}, V_{th0}, E_{ta0}, V_{dd}, and other primary parameters, robust circuit design research under different conditions is fully supported.

PTM models gate tunneling leakage relying on scalable models of leakage current. Calibration with published 65 and 45 nm data has shown a reduction by about 25–1,000 times in gate tunneling leakage for the same EOT. It has been shown that HK-MG technology will not only suppress gate leakage, but also boost driving current significantly. Figure 9.10 shows the smooth predictions of I_{on} and I_{off} at the 32 nm node with and without HK-MG for all three V_{th} processes. I_{off} of high V_{th} deviates from the nominal trend due to the GIDL and tunneling current. Besides the prediction of I-V, the scaling trends of gate and parasitic capacitances are covered in PTM, since they are important for dynamic circuit performance. PTM validation shows a smooth prediction of both speed and power consumption from the 65 nm node down to the 32 nm node.

Li et al. [5] have proposed a predictive strategy for simultaneous exploration of low-power CMOS process and design concepts for 22 nm low-power designs. Authors have evaluated critical performance metrics, e.g., speed and power, with various technological components and design choices with scaled CMOS and have incorporated the general PTM methodology, with customised enhancements of transistor-level and interconnect-level physical

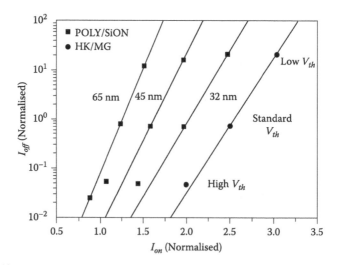

FIGURE 9.10
I_{on} and I_{off} predictions at the 32 nm node for various process choices. (After Zhao, W., Predictive Technology Modelling for Scaled CMOS, PhD thesis, Arizona State University, 2009.)

effects. The customised PTM models have been calibrated to 90 to 45 nm Poly/SiON silicon and high-k/metal gate (HK-MG) data.

9.2.1 PTM for FinFET

Beyond the 22 nm technology node, more radical solutions will be necessary to meet the scaling criteria for off-state leakage. The double-gate MOSFET (DG) or FinFET is regarded as a promising alternative device for the nanoscale design because of its improved scalability and the effective suppression of short-channel effects. When the body silicon thickness (T_{Si}) is sufficiently thinner than the channel length, short-channel effects, such as V_{th} lowering, DIBL, and increased subthreshold swing, can be effectively suppressed. With a lightly doped channel, the threshold voltage of a FinFET transistor is weakly affected by random dopant fluctuations. The FinFET device is electrostatically more robust than bulk CMOS since two gates are used to control the channel. The front and back gates can be connected together or biased independently, using the front gate to switch the transistor on/off and the back gate as a control signal. At the 32 nm node, it may improve the I_{on}/I_{off} ratio by more than 100%. About 20 sets of published I-V data from the 250 nm node to the 45 nm node at room temperature were used to verify the PTM for FinFETs. By tuning 10 primary parameters, the predicted I-V characteristics are compared for verification. Figure 9.11 demonstrates the matching of a FinFET transistor with $L_{eff} = 30$ nm.

9.3 Process-Aware Design for Manufacturing

As the CMOS technology continues to scale down in the sub-100 nm regime, power dissipation and robustness of a circuit with respect to process variations pose major design challenges. Variability arising from advanced silicon technologies, such as strain engineering, is increasingly affecting the circuit performance. The control of process fluctuations has not kept pace with rapidly shrinking device dimensions. It is important to characterise and quantify systematic, random, die-to-die, and within-die transistor variability in order to control variability from both the manufacturing and the design angle. Variation-aware statistical analysis technologies are needed to explore and optimise the process and design methodologies. The Paramos tool from Synopsys links SPICE models directly to manufacturing conditions by extracting process-aware SPICE compact models that combine calibrated TCAD simulations with global SPICE extraction. It allows users to simulate the impact of process variability (statistical or systematic) on circuit performance. This methodology also provides a physically based variation model for statistical timing simulations of circuit performance, allowing

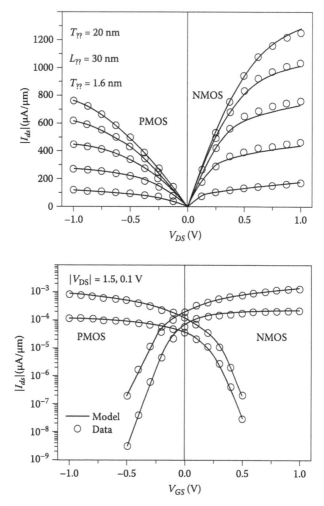

FIGURE 9.11
Verification of FinFET in PTM. (After Zhao, W., Predictive Technology Modelling for Scaled CMOS, PhD thesis, Arizona State University, 2009.)

them to explore a design's sensitivity to real physical process parameters. Seismos and Paramos from Synopsys address two major sources of variability in a design: proximity variations caused by stress and other neighbourhood effects and global variations due to the spread of manufacturing process parameters across different die and wafers.

Designing circuits with high yield under parameter variations has emerged as a serious challenge. The process-aware SPICE model is a deterministic model that includes systematic and random process variations and improves the understanding of the process variability impact design.

Other important features of the process-aware SPICE model are that it is measurable, adjustable, and statistically independent, can be monitored and recorded as part of an ongoing manufacturing process, allows desensitisation of design to process variations and design-specific process centring, and enables engineers to consistently optimise the process and design with minimal experimental efforts, which results in significant productivity improvements.

Process-aware design for manufacturing involves analyses of variability effects at the custom/analogue design stage that enable the designers to see how much they can push the design rules and realise the full potential of technology scaling. The process-dependent SPICE models allow direct access to process parameter variations in circuit design [6]. For example, change of the gate length results in change of the device parasitic, which can be included in timing analysis by the circuit designer for examining the delays leading to variation-aware circuit design. The use of PCM significantly improves design for manufacturing (DFM) by allowing for accurate design sensitivity analysis and parametric yield assessment, as a function of statistically independent and measurable process variations.

9.4 Process Compact Model

In the following, we present a simulation methodology for strain-engineered MOSFETs that allow the flow of pertinent information between process and design engineers without the need for disclosing the detail of process technology. Compact SPICE model parameters are obtained using parameter extraction strategy by using a polynomial function of process parameter variations. A strategy to acquire compact SPICE model cards has been developed. As a case study, SPICE models are used to identify the impacts of process variability on the performance of inverter circuits with strain-engineered MOSFETs.

Technology CAD (TCAD) is a powerful tool to identify the root causes for yield loss and is used to study device sensitivities on process variations. Currently, TCAD is heavily used in device research and process integration phases of technology development. However, a major trend in the industry is to apply TCAD tools far beyond the integration phase into manufacturing and yield optimisation. Linking of process parameter variations (via design of experiments) with the electrical parameters of a device through a process compact model (PCM) is discussed. Application of stable and well-calibrated TCAD tools as an aid for manufacturing of process-induced strain-engineered CMOS is described.

Toward extended TCAD, in process modelling, generally a systematic design of experiments (DoE) run is performed. DoE experimentation is systematically set up to study the control over process parameters and arbitrary choice of device performance characteristics. The models developed from DoE are known as process compact models (PCMs), which are analogous to compact models for semiconductor devices and circuits. PCMs may be used to capture the nonlinear behaviour and multiparameter interactions of manufacturing processes [7, 8]. SPICE process compact models (SPCMs) can be considered an extension of PCMs applied to SPICE parameters. By combining calibrated TCAD simulations with global SPICE extraction strategy, it is possible to create self-consistent process-dependent compact SPICE models, with process parameter variations as explicit variables. This methodology brings manufacturing to design, so that measurable process variations can be fed into design [9]. To design robust circuits using strain-engineered MOSFETs, the effect of process variability on the circuit model parameters is examined.

The process compact model methodology consists of TCAD simulations, using the process and device models that are calibrated to strained Si and process-dependent compact SPICE model extraction (Figure 9.12). The parameter extraction is performed using the parameter extraction tool Paramos [10], which interfaces TCAD or experimental data and directly generates process-aware SPICE models. The process-aware SPICE models allow designers to account for process variability and develop more robust designs.

The process compact model generation strategy includes the following: (1) Capture the process-device relationships between the process parameters and device performance of a semiconductor manufacturing process. (2) PCM is robust, fast to evaluate, and can be embedded into other environments such as PCM Studio, spreadsheet applications, and yield management

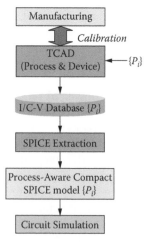

FIGURE 9.12
Compact SPICE model extraction and validation methodology.

systems. (3) PCM is analogous to device compact models, which capture electrical behaviour and can be derived from measurements or simulations.

SPICE process compact models (SPCMs) can be considered an extension of PCMs applied to SPICE parameters. Using a global extraction strategy, available from the Synopsys tool Paramos, pertinent compact SPICE model parameters are simultaneously obtained as a polynomial function of process parameter variations. The extraction procedure is performed using Paramos, which will deliver an Extensive Markup Language (XML) file containing the extracted SPICE model parameters. This methodology brings manufacturing to design, so that measurable process variations can be fed into design. Additionally, design sensitivity to process can be fed back to manufacturing so that product-dependent process controls can be performed. Here, the chosen SPICE model parameters (Y_i) are extracted as an explicit polynomial function of normalised process parameter variations ($\tilde{P}_j$), as shown in Equation (9.1). Process parameter variations are normalised with respect to the corresponding standard deviation of the parameter, as shown in Equation (9.2). Such a normalisation process enables the encryption of proprietary information like the absolute values of the process parameters.

$$y_i = y_i^0 + \sum_j \sum_{n=1}^{N} a_{ij}^n \tilde{p}_j^n \tag{9.1}$$

where Y_i is the nominal value of the ith model parameter, j is the jth process parameter, N is the highest order of polynomial, a_{ij}^n is the process coefficient of the jth process parameter for the ith SPICE model parameter, and for order n of the polynomial, $\tilde{p}_j$ is the normalised process parameter, defined as

$$\tilde{p}_j = \frac{p_j - p_j^0}{\sigma_j} \tag{9.2}$$

where p_j is the jth value of the process parameter, p_j^0 is the nominal value of the jth process parameter, and σ_j is the standard deviation of the jth process parameter. In our study, we used BSIM4 SPICE model parameters as a quadratic function of process parameters. This model is easily scalable to higher orders of polynomial (N) for higher accuracy of extraction [11]. The current extraction strategy of the SPICE model parameters involves extraction of nominal SPICE parameters $\left(y_0^i\right)$, followed by extraction of process coefficients $\left(a_{ij}^n\right)$ and reoptimised nominal values of SPICE parameters $\left(y_0^i\right)$. In the following, we discuss the generation of process compact models (PCMs). PCM Studio offers an accommodating front end to construct polynomial, Hermite polynomial, and neural network PCMs, based on TCAD simulation data. Creating SPICE process compact models (SPCMs) is not directly supported, though there is a plug-in for PCM Studio that simplifies the construction of input files for the Paramos extraction engine, which is used to build SPCMs.

9.4.1 PCM Analysis

Process compact models can be used with various kinds of data to perform different types of analysis. Possible data sources include in-process measurements, electrical test data, nominal process conditions, electrical target specifications, and random values. Numerical optimisation allows the use of response (device characteristics) as input in order to obtain estimated values for process parameters [10]. The possible analyses are:

1. PCM evaluation. This type of analysis uses in-process measurements, nominals, or randomly generated values for the process parameters to evaluate the PCM and generate device characteristics. This is a basic analysis and requires no numeric optimisation.

2. Reverse analysis. Reverse analysis estimates the distribution of certain nonmeasurable process parameters based on data for the rest of the parameters and electrical measurements for the device characteristics.

3. Feed-forward analysis. Feed-forward analysis estimates the distribution of critical parameters (to understand the amount of control required) based on data or nominals for the rest of the process parameters and target device specifications (for the responses). Technically, this corresponds to a reverse analysis with a fixed target value for the responses. Feed-forward analysis does not support SPCMs.

All three analyses have a commonality: they use data from measurements or specifications to estimate process parameters or device characteristics that are difficult or expensive to measure. Conceptually, there may be two different cases: (1) The values for all the parameters are available, and those for the responses should be estimated as illustrated in Figure 9.13 for the case of a PCM evaluation. (2) The values for a part of the parameters and responses are available, and the corresponding values for the rest of the parameters should be estimated. This is illustrated in Figure 9.14 for reverse analysis.

FIGURE 9.13
Schematic view of PCM evaluation.

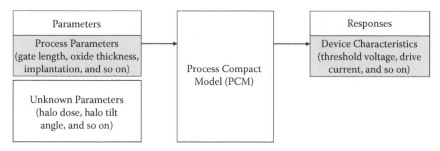

FIGURE 9.14
Schematic view of reverse analysis.

9.5 Process-Aware SPICE Parameter Extraction

To extract the model parameters, process and device simulations were first performed using typical CMOS process flow. The model parameters extracted are for the nominal process conditions and various drawn gate lengths. One of the SPICE parameters, namely, voltage (V_{th}), as a function of process parameters, has been extracted. In order to validate the compact SPICE model, for a given set of process conditions and device bias states, I-V curves obtained from TCAD simulations are compared with those obtained from Paramos using a process-dependent compact SPICE model card. Figure 9.15 shows the current-voltage characteristics for a 45 nm n-MOSFET. The dots show the TCAD simulation data, and the solid lines show the electrical characteristics generated by a global SPICE model.

The compact model parameters for the CMOS devices are extracted using the BSIM4 MOSFET model. The process-aware model parameters are extracted from electrical simulations where process parameters such as gate length, gate oxide thickness, halo dose, extension dose, and rapid thermal annealing (RTA) were varied. These parameters are selected to model the process variability because of their primary impact on the electrical characteristics of the device. Table 9.1 summarises the parameters and ranges chosen for the 45 nm CMOS process optimisation. The influence of process variation on threshold voltage for p- and n-MOSFETs has been studied.

The process compact model (PCM) is validated by examining the fit of simulation using Hermite polynomial or neural network models. The fits for the threshold voltage (V_{th}) for p- and n-MOSFETs are shown in Figure 9.16(a) and (b), respectively.

As an example, we have chosen the SPICE model parameter threshold voltage (V_{th}) extracted as an explicit polynomial function of normalised process parameter variations (P_i^n), as shown in Equation (9.3):

$$V_{th} = V_{th0} + \sum \sum a_i^n P_i^n \tag{9.3}$$

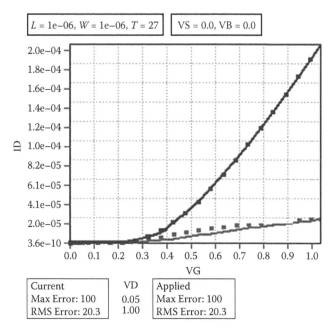

FIGURE 9.15
Current-voltage characteristics for n-MOSFETs with 45 nm gate length.

where V_{th0} is the nominal value of threshold voltage, i is the ith value of the process parameter, a_i^n is the process coefficient of the ith process parameter for the SPICE model parameter and for order n of the polynomial, and P is the normalised process parameter. Such a normalisation process (P) enables the encryption of proprietary information like the absolute values of the process parameters. In our work, we consider the BSIM4 SPICE model parameters as quadratic functions of process parameters. This model is easily scalable to higher orders of polynomial (n) for higher accuracy of extraction. The SPICE model parameter, such as threshold voltage (V_{th}), involves extraction of nominal SPICE parameters (V_{th0}), followed by extraction of process coefficients a_i^n

TABLE 9.1

Process Parameters under Study and the Corresponding
Allowed Variation

Parameter	Parameter Name	% Variation
Gate length	L_g	±30%
Gate oxide	G_{ox}	±20%
Halo implant dose	Halo_Dose	±25%
Extension implant dose	Ext_Dose	±10%
Peak temperature for RTA	RTA	±10%

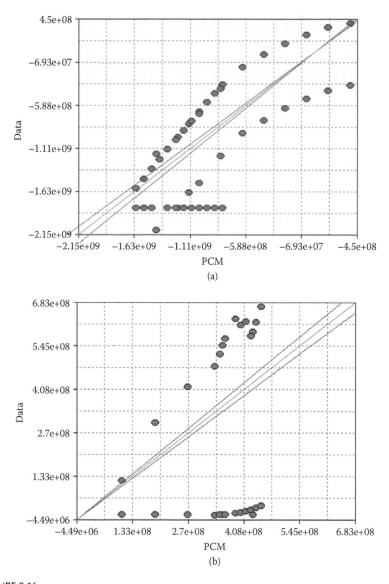

FIGURE 9.16
(a) PCM validation for p-MOSFETs with respect to TCAD simulation. (b) PCM validation for n-MOSFETs with respect to TCAD simulation.

and reoptimised nominal values of SPICE parameters (V_{th0}). Equations (9.4) and (9.6) show one of the SPICE parameters, V_{th}, as a function of process parameters for process-induced strained Si p- and n-MOSFETs, respectively. The threshold voltage model for strain-engineered p-MOSFETs is obtained using a first-order polynomial as a function of gate length (L_g) and germanium mole fraction (Ge):

$$V_{th} = -V_{th0} - (L_g - \alpha_1).\beta_1 - (Ge - \alpha_2).\beta_2 \tag{9.4}$$

where the coefficients α_1, α_2, β_1, and β_2 can be calculated from TCAD simulation or experimental data. The expression for V_{th} obtained is given as

$$V_{th} = -0.3435 + (Lg - 85)/40* (-0.0824303)$$
$$+ (Ge - 875005e + 21)/6.25005e + 21* (-0.0172263) \tag{9.5}$$

Threshold voltage models for strain-engineered n-MOSFETs have been obtained using first-order polynomials as a function of gate length (L_g) and nitride cap layer thickness (T_{SiN}):

$$V_{th} = V_{th0} + (L_g - \alpha_1).\beta_1 + (T_{SiN} - \alpha_2).\beta_2 \tag{9.6}$$

and the corresponding threshold voltage expression is given by

$$V_{th} = 0.215441 + (Lg - 80)/45*0.0780802$$
$$+ (SiN - 0.065)/0.02*0.0148119 \tag{9.7}$$

Here, SPICE parameters are represented as first-order polynomial functions of process parameter variations. The threshold voltage parameter generated by the global SPICE model shows the maximum error is approximately 10% and the root mean square (RMS) error is approximately 4%. These results show that the global model can be used to predict the electrical behaviour of the devices in the absence of process variability. Figure 9.16(a) and (b) clearly indicates that the process-aware model developed above can account for process variability-induced performance variation.

9.5.1 Circuit Modelling

As a case study, simple digital circuits are simulated to assess the accuracy of the extracted circuit model parameters. Circuit simulations are first performed using the TCAD data. The inverter circuit is simulated using the devices created with the 45 nm process flow by mixed-mode TCAD circuit

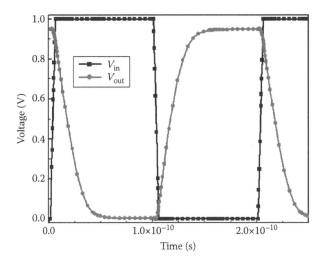

FIGURE 9.17
Switching characteristics for the PSS-CMOS inverter.

simulations using Sentaurus Device. The TCAD simulations are compared with the HSPICE simulations using the process-aware model. In the case of the circuit simulations performed using Paramos and the process-aware model parameters, a stress-dependent threshold voltage is added to account for the circuit performance. Switching characteristics using a process-aware model for the PSS-CMOS inverter is shown in Figure 9.17, where input voltage and output voltage flowing through the n-MOSFETs are illustrated. The results from simulations are also compared to SPICE simulations performed using the process-aware model parameters [10, 12]. In addition, the absolute model error is calculated using

$$Error = \frac{Q_{SPICE} - Q_{TCAD}}{\sqrt{\Sigma(Q_{TCAD})^2 / N}} \qquad (9.8)$$

which indicates the model accuracy. These studies demonstrate the accuracy and robustness of the process-aware circuit model parameters and their suitability for possible applications in strain-engineered CMOS standard cells, as well as more complex circuit elements.

9.6 Summary

In this chapter, the predictive technology model of MOSFETs, which offers a generic, open-source tool for early-stage design research, has been introduced. A systematic study based on technology CAD is taken up for the design and virtual wafer fabrication (VWF) of strain-engineered MOSFETs in Si CMOS technology. A manufacturable process has been considered to induce uniaxial stress in the channel region to obtain enhanced CMOS performance. A methodology for capturing process variability in SPICE models has been presented. Process parameters considered are gate length (L_g), gate oxidation temperature, halo dose, cap layer thickness, Ge mole fraction, and V_{th} implant dose. The methodology involves global extraction of process-dependent SPICE model parameters from strained Si calibrated TCAD simulations. The model is validated by comparing device characteristics from the extracted SPICE parameters with those from TCAD simulations. The SPICE parameters are used to identify the impact of process variability on critical circuit performance. The extracted models are employed in an inverter circuit with strain-engineered MOSFETs. The process-dependent SPICE models extracted from Paramos provide a key bidirectional link between the variability of the manufacturing process and circuit simulations of chip performance.

Review Questions

1. Compare physically based, semiempirical, empirical, and compact models.
2. Describe briefly the applications of the SPICE simulation tool.
3. What is the predictive technology model?
4. Describe briefly the applications of PTM.
5. What is meant by process compact SPICE model?
6. Compare the process compact model with the standard SPICE model.
7. What are the latest versions of the process compact SPICE model and standard SPICE model?
8. Describe the PCM methodology.
9. Write the steps to be followed to determine the process-dependent SPICE parameters for the process parameter variations, such as gate critical dimension, annealing, and implantation.
10. Describe briefly: (a) worst-case corner models, (b) statistical corner models, and (c) TCAD-based corner models.
11. Describe the predictability of the BSIM model for the process and layout variation.
12. What are the impacts of PCM for circuit-level analysis?

References

1. Y. Cao, *Predictive Technology Model for Robust Nanoelectronic Design*, Springer Science + Business Media, LLC, New York, USA, 2011.
2. W. Zhao, Predictive Technology Modelling for Scaled CMOS, PhD thesis, Arizona State University, 2009.
3. W. Zhao and Y. Cao, New Generation of Predictive Technology Model for Sub-45 nm Early Design Exploration, *IEEE Trans. Electron Dev.*, 53, 2816–2823, 2006.
4. *BSIM4 User Guide*, University of California, Berkeley, 2008.
5. X. Li, W. Zhao, Y. Cao, Z. Zhu, J. Song, D. Bang, C.-C. Wang, S. H. Kang, J. Wang, M. Nowak, and N. Yu, Pathfinding for 22nm CMOS Designs Using Predictive Technology Models, *Proc. Custom Integrated Circuits Conference (CICC'09)*, 227–230, 2009.
6. S. Krishnamurthy, V. K. Dasarapu, Y. Mahotin, R. Ryles, F. Roger, S. Uppal, P. Mukherjee, A. Cuthbertson, and X.-W. Lin, Si-Based Process Aware SPICE Models for Statistical Circuit Analysis, *Proc. Nanotechnol.*, 3, 897–900, 2008.
7. T. K. Maiti, S. S. Mahato, P. Chakraborty, S. K. Sarkar, and C. K. Maiti, Scaling of Strain-Engineered MOSFETs, *IETE J. Res.*, 53, 263–236, 2007.
8. C. K. Maiti, T. K. Maiti, and S. S. Mahato, Strain-Engineered MOSFETs, *Semiconductor India Mag.*, June 2008.
9. R. Borges, T. Ma, W.-C. Ng, S. Krishnamurthy, and L. Bomholt, Implementation of TCAD-for-Manufacturing Methodology Using Process Compact Models, *Proceedings of International Conference on Solid-State and Integrated Circuit Technology (ICSI'06)*, 2006.
10. Synopsys, Inc., *Paramos User Guide*, version B-2008.06, Mountain View, CA, June 2008.
11. S. Tirumala, Y. Mahotin, X. Lin, V. Moroz, L. Smith, S. Krishnamurthy, L. Bomholt, and D. Pramanik, Bringing Manufacturing into Design via Process-Dependent SPICE Models, *Proc. Int. Symp. Qual. Electronic Design (ISQED'06)*, 806–810, 2006.
12. Synopsys, Inc., *HSPICE User Guide: Simulation and Analysis*, version C-2009.03, Mountain View, CA, 2009.

10

Process-Aware Design of Strain-Engineered MOSFETs

The decrease in device size or scaling into deep submicron feature sizes has introduced many design challenges that did not exist before or many of which could be ignored. Some of the deep submicron issues are exponential increase in leakage power, thermal issues and hot spots on the chip due to increasing transistor density, deterioration in reliability due to increase in various types of noise (e.g., cross-coupling noise, power grid noise), soft errors due to cosmic radiation and continuous scaling of supply voltage, and fabrication defects. Advanced semiconductor manufacturing technology demands techniques for efficiently designing high-performance, low-power integrated circuits, with shorter time-to-market design time [1]. It is necessary to link manufacturing variation information back to design, enabling custom integrated circuit (IC) designers to optimise layouts and maximise yields. The main challenges in manufacturing are reducing cycle time, enhancing production quality and variability control, improving equipment productivity, reducing the environmental impact, supporting heterogeneous integration, advancing system integration, and functionalising packaging. The merging of "beyond CMOS" and advanced "more than Moore" devices and processes to create a complementary metal-oxide-semiconductor (CMOS) backbone will further increase process variability and other reliability issues. Thus, the need for new device/circuit architectures, metrology, and characterisation techniques will increase.

Process variations refer to those variations caused due to the imperfections in different steps of the IC manufacturing process. One such example is the limited resolution of the photolithography in the fabrication process, which results in variations in the width and length of transistors. The aggressive scaling of silicon technology has enabled dramatic improvements in integrated circuit performance. However, the control of semiconductor manufacturing processes has become increasingly difficult and expensive. Process variations could be identified at different levels depending on various stages of the manufacturing process: wafer to wafer, die to die, and within die. Process variations have an overall unpredictable nature because of not having enough control over different steps of the fabrication process.

As the technology advances, the critical dimensions (CDs) of many layers are well into the subnanometer regime, and film thickness has approached atomic layer dimensions [2]. Beyond 22 nm devices, advanced components with lower scaling factors, including non-CMOS devices, are expected. This results in a greater variability. For example, the gate dielectric of a typical 65 nm node is on the order of four atomic layers thick. It would be impossible for any process to place four atomic layers precisely on the gate oxide of every transistor on an entire wafer. Also, with gate length CDs in the sub-40 nm range, 4 nm of variability would represent a 10% change in the CDs [3]. As variability in manufacturing processes has grown more severe, variations in device parameter values have grown in proportion to nominal values. In turn, wider distributions for device parameters have led to increased variability in circuit performance, causing worsened yield degradation in successive technology generations.

Technology computer-aided design (TCAD) tools can simulate IC fabrication process technology and device characteristics and are indispensable for advanced technology development and manufacturing. TCAD is now an integral part of integrated circuit manufacturing due to its predictive capability for the process, device, and circuit simulations. TCAD also has the power to analyse accurately the impact of process parameter variations on device characteristics and may be used to address and control process variability as needed for modelling the semiconductor manufacturing process. In this chapter, a systematic study based on technology CAD is taken up for the optimisation of strain-engineered metal-oxide-semiconductor field-effect transistors (MOSFETs) in Si CMOS technology via virtual wafer fabrication (VWF). A simple and manufacturable process recipe is developed to induce uniaxial stress in the channel region to obtain enhanced performance in the CMOS.

10.1 Process Design Co-Optimisation

The general flow for fabrication of integrated circuits is comprised of several steps. In the front-end processes, dopants are implanted and diffused into the silicon substrate and various materials are repeatedly deposited and patterned to build active devices such as MOS transistors. In back-end processes, layers of interconnects used as wires between active devices are created using successive repetitions of deposition, patterning, and polishing. Variability in circuit performance is a rapidly growing concern in the semiconductor industry, and a potential roadblock in circuit design. To avoid the negative impact of manufacturing variations on circuit performance, two approaches are being taken. The first approach is to apply a renewed focus on process control from a manufacturing perspective, in an effort to directly reduce the variations in device parameters. The second approach comes from the design perspective, where practises can be developed to decrease circuit sensitivity to process variation. Both approaches rely on exploration through the use of simulation frameworks that capture the detailed interaction between manufacturing variation and the resulting circuit performance variability. With such a framework, one can determine the most deleterious sources of device parameter variation, and then identify the effects of a certain flavour of process control, or search for sensitivity-reducing design techniques.

Yield and performance are the foremost concerns for device design in the semiconductor industry, and a clear understanding of their sensitivity to process parameters is the key for better control. In this chapter, we shall discuss a TCAD methodology that addresses the manufacturing challenges posed by rising technological complexity, increasing process variability, and shrinking time-to-market windows. Using TCAD process and device simulations for typical CMOS technology as input, process compact models (PCMs) are created to enable efficient analysis of complex and multivariate process-device relationships. PCMs are then applied to enhance manufacturability and process control. A yield optimisation technique is also proposed to suppress the variability of a device optimised for subthreshold operation. The goal of this technique is to construct and inscribe a maximum yield region composed of oxide thickness, gate length, cap layer thickness, Ge mole fraction, and channel doping concentration. The centre of this cube is chosen as the maximum yield design point with the highest immunity against variations. By using the technique, a transistor is optimised to design an inverter circuit.

10.2 Classifications of Variation

Circuit parametric variations arise either from fluctuations in the wafer manufacturing process, known as intrinsic variation, or from the dynamic operation of the circuit, for example, local temperature and supply voltage variations, known as extrinsic variation. For this discussion, we are concerned with sources of intrinsic variation, although comprehensive models for intrinsic variation allow circuit simulators to account for extrinsic variation. Figure 10.1 is a chart showing the possible sources of yield loss. Yet, except for the random defects component and the physics component (such as stress, electromigration, and reliability), it could have easily been used as a map of the sources of variability.

Variability occurs as a function of location (spatial) or time (temporal). Spatial variability can be grouped based on its inherent length scale. Long-length scales, for example, a cross-wafer nonuniformity in etch and deposition processes, lead to interdie variation. On the other hand, short-length

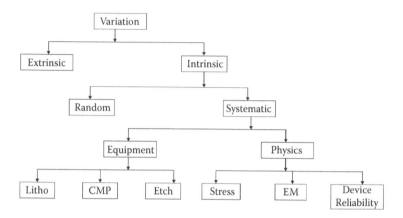

FIGURE 10.1
A comprehensive chart of the sources of yield loss.

scales, for example, optical interference effects, result in intradie variation. In advanced CMOS technologies, intradie variation is comparable to and in some cases significantly larger than interdie variation. A further classification distinguishes between systematic variation and random variation.

Systematic variation is attributed to known and predictable phenomena. Examples include optical proximity effects, layout-induced stress, and the well proximity effect. Optical proximity effects have received a great deal of attention in recent years, with optical proximity emerging as a common technique to mitigate these effects. The partial compensation of systematic variation in the process and design phases afforded by these types of tools provides more tolerance for technology scaling. Random variation, on the other hand, is due to inherently unpredictable fluctuations of the manufacturing process. It cannot be controlled with standard statistical process control (SPC) techniques, making it difficult or expensive to minimise. Examples include fluctuations in channel doping, gate oxide thickness, and gate length. Some studies indicate random variation could have the largest impact on chip yield in future, highly scaled, CMOS technologies. Manufacturing-induced variation can be thoroughly characterised and decomposed into several deterministic components, including wafer-to-wafer variation, across-wafer variation, within-die variation, and pattern-dependent variation. Wafer-to-wafer variation arises from variation in the state of the manufacturing tool. Over time, the conditions of a given process may drift so that wafers passing through a given process step near the beginning of a lot undergo a slightly different process than those processed near the end. Typically sources of wafer-to-wafer variation exist in process steps, such as the etch and chemical mechanical polishing (CMP) modules; however, due to major advances in factory quality control, principally through automated statistical process control, wafer-to-wafer variation can be considered a lesser source of concern than across-wafer variation.

10.3 Designs for Manufacturing and Yield Optimisation

In the following, we discuss a general formulation of the device optimisation problem that is composed of the selected device design parameters and constraints. Thereafter the idea behind the yield maximisation process is established, and the problem is formalised by the yield maximisation technique.

The approach [4] adapted in subthreshold transistor design consists of exploiting a 3D parameter design space, constructed by T_{ox}, L_g, and N_{halo}. Hence, the yield optimisation problem is as follows:

$$\text{Given: } \sigma_{T_{ox}}, \sigma_{L_g}, \sigma_{N_{halo}},$$

$$\max_{x \in R^3} \quad Yield = P\{C(x) = 1\} \tag{10.1}$$

where $x = [T_{ox}, L_g, N_{halo}]$ is the set of design variables, σ_i is the standard deviation of the ith design parameter, and $C(x)$ is a Boolean random variable function, defined by the bounds of the critical delay (I_{onmax}) and the maximum threshold voltage (V_{tmax}). $C(x)$ is formulated as

$$C(x) = (I_{on}(x) \leq I_{on\max}) \quad and \quad (V_t(x) \leq V_{t\max}) \tag{10.2}$$

Therefore, $P\{C(x) = 1\}$ is the probability that a device $x = (T_{ox}, L_g, N_{halo})$ meets the performance and power constraints in the presence of variations in the design parameters.

To solve the optimisation problem, Equation (10.1), the first step is to find a 3D space, generated by the three device design parameters, bound by the power and performance constraints. This space is called the feasible region, F_c. In addition, an estimate of the probability of placing a device in F_c should be calculated, that is, the probability that a device $x_i = (T_{oxi}, L_{gi})$ can satisfy the desired constraints, on-state drive current (I_{on}), and threshold voltage (V_t). To estimate such a probability, $P\{C(x_i) = 1\}$, a cube is formed in the 3D parameter design space, where all points within the cube satisfy the constraints. For clarification, a similar problem with two design variables T_{ox} and L_g is denoted in Figure 10.2. Any point inside this plane represents the construction strain-engineered MOSFET dimensions, corresponding to the respective ordered pair (T_{ox}, L_g). A feasible region is defined in terms of the problem constraints. Any device x_i above the V_t curve in Figure 10.2 satisfies the power constraint, and any device below the I_{onmax} curve meets the performance constraint. Therefore, all the devices lying in the intersection of the defined zones can satisfy both constraints, as depicted by the shaded region in Figure 10.2.

For the last constraint, the yield maximisation problem is reduced to an inscribed rectangle that is formed by four corner devices: (T^l_{ox}, L^l_g), (T^l_{ox}, L^u_g), (T^u_{ox}, L^l_g), and (T^u_{ox}, L^u_g) in the 2D feasible region. The centre of the maximum

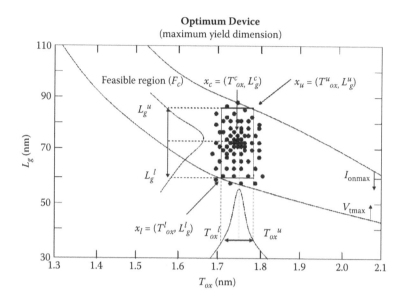

FIGURE 10.2
Simplified problem in 2D.

yield rectangle $x^c = (T^c_{ox}, L^c_g)$ represents a device with the set of design values most immune to the variations. Finally, technology CAD (TCAD) simulations are carried out to verify the optimal design (x^c) yield, which is defined as the percentage of the total devices (scattered points) whose I_{on} and V_t values fall within the feasible region F_c.

10.3.1 Process Optimisation

Process variability has become a primary concern with regard to manufacturability and yield [5]. As device dimensions shrink, the sensitivity of device performance to process variation also increases. With 45 nm processes, it is imperative to develop a systematic TCAD-based methodology to design, characterise, and optimise manufacturability to increase yield [6]. As the manufacturability of a process technology may be evaluated by the process window, defined as the area between the lower and upper limits of the critical process variables that yield acceptable device performance, in the following, we use the Sentaurus PCM Studio for the strain-engineered MOSFETs.

10.3.2 Process Parameterisation

To demonstrate process optimisation using PCM Studio, one device parameter, e.g., threshold voltage (V_t), is chosen and the process is optimised with respect to V_t. As an example, we optimise the device performance by minimising threshold voltage (V_t), which mainly depends on

TABLE 10.1

Process Variability and Range

Parameter	Parameter Name	% Variation
Gate length	L_g	±20%
Gate oxide	G_{ox}	±10%
Halo implant dose	Halo_Dose	±25%
Extension implant dose	Ext_Dose	±15%
Peak temperature for RTA	RTA	±10%

the following parameters: halo implant dose (Halo_Dose) and extension implant dose (Ext_Dose), gate length (L_g), gate oxide thickness (G_{ox}), and peak temperature for rapid thermal annealing (RTA), which modifies the doping concentration in the channel region. The optimisation problem consists of finding the best combination of the above parameters that produces the desired threshold voltage. The visual optimisation procedure [7] allows one to put constraints on the input parameters, which, however, are motivated by the manufacturing considerations. For an example, we may set a minimum for the gate length to obtain a nominal threshold voltage. Table 10.1 summarises the parameters and ranges chosen to optimise strain-engineered MOSFETs.

10.3.3 Smoothness and Sensitivity Analysis

For the determination of the influence of tolerances in the technology process optimisation, different process parameters have been varied. Before running the systematic TCAD simulations, smoothness and sensitivity analysis is performed to determine the critical process variables and suitable ranges for the experimental design. Input parameters are varied one at a time, within the previously specified range. The effort of this one-at-a-time parametric variation grows linearly with the number of input parameters, which allows us to potentially examine many parameters. While three points per parameter are, in principle, sufficient to capture second-order effects, they are insufficient to assess whether some variation is truly physical or is caused by simulation artifacts such as meshing noise. Five to 10 points is a better choice and also indicates which order of design of experiments (DoE) to use. The effort remains reasonable. This analysis also characterises the sensitivity of the nominal device to each parameter and helps to select the parameters to be used in the computationally more expensive PCM. Figure 10.3(a) and (b) illustrates the sensitivity of the p- and n-MOSFET responses with respect to the halo implant dose, respectively.

Figure 10.4 shows a normalised histogram plot summarising the sensitivity analysis for the critical process steps. The variation of each output parameter for the specified input range is normalised to the maximum value; that is, the y axis range is 0 to 1.

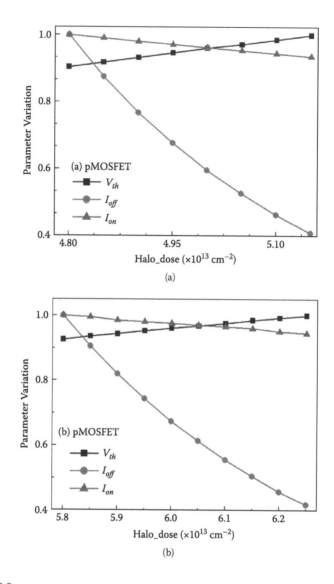

FIGURE 10.3
(a) Sensitivity response of p-MOSFET with respect to halo implant dose. (b) Sensitivity response of n-MOSFET with respect to halo implant dose.

10.3.4 Visual Optimisation

A very useful way of visualising and analysing PCM-generated data is the parallel coordinate plot [7], which plots multivariate data in a single representation. It is created by mapping coordinates in a multidimensional space

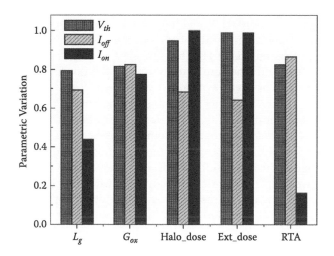

FIGURE 10.4
Sensitivity analysis of process variability for 45 nm process-induced strained Si MOSFETs.

onto a set of parallel axes, one for each input and output parameter. A line connects the corresponding coordinates. In particular, the parallel coordinate plot is an ideal environment in which to perform a visual optimisation [8]. Automatic optimisation procedures, regardless of the specific optimisation algorithm, suffer from the fact that all optimisation criteria must be included in a single "fitness function," which is subsequently maximised. In practise, the different optimisation criteria flow into the fitness functions through relative weighting. However, since it is usually impossible to meet all optimisation targets perfectly, it is difficult to select the appropriate weighting. Often, the weighting can be determined only in hindsight, after the capabilities of the technology are measured against the requirements. The combination of fast PCMs in a visual, interactive environment addresses many of these issues. Using the PCM Studio, we have generated a large number of experiments, for example, 1,200, with uniform random distributions for each input parameter. This is similar to creating an initial population for a genetic algorithm. We continue with the visual optimisation by highlighting process-induced strained Si p- and n-MOSFETs examples.

10.4 Performance Optimisation

The combination of SProcess, SDevice, PCM Studio, and Sentaurus Workbench forms a powerful design for manufacturing (DFM) TCAD environment. In this study, a total of 1,200 experiments were generated. The process and device simulation results are subsequently used as the basis for generating a process compact model (PCM), which encapsulates the relationships between input (design) and output parameters. The PCM automatically correlates design parameters to the tolerances. The ranges are normalised to 1 (see Figure 10.4), with the centre representing the nominal value for each parameter. The process was optimised with respect to threshold voltage, channel stress, device current, and transconductance. Parallel coordinate plots link the simulation results to the design variation. The parameter values and ranges indicate whether the domain has been covered sufficiently. The yellow region is the constraint of the parameters and the outputs that satisfy the range of design specifications. Red lines within this region depict the successful design. For the case study of p-MOSFET threshold voltage optimisation, we allowed a threshold voltage variation within 0.007–0.243 V. We put a variation limit on gate length by narrowing the experiment selection, resulting in a 5% lower V_t compared to the nominal value. We select only lower V_t, which means reducing on-state voltage.

The optimisation procedure is continued, and finally we perform a further screening on a germanium mole fraction for process-induced strain-engineered p-MOSFETs and nitride cap layer for process-induced strain-engineered n-MOSFET, resulting in a combination that gives a Ge mole fraction and SiN thickness, generating the optimised V_t. By repeating the above optimisation procedure, the device performance may further be improved to obtain V_t within 1%. The process conditions satisfying the specifications for V_t indicated by black lines in the parallel coordinate plot provide information about how well the domain space is covered with the chosen DoE. Figure 10.5(a) and (b) shows the process compact model evaluation scenarios for process-induced strain-engineered p- and n-MOSFETs, respectively.

Figure 10.5(a) is a parallel coordinate plot that links the simulation results to the design variation of the gate length and germanium mole fraction (Ge) in the embedded SiGe source/drain region for process-induced strain-engineered p-MOSFETs. Similarly, Figure 10.5(b) is a parallel coordinate plot that links the simulation results to the design variation of the gate length and cap layer thickness (SiN) for process-induced strain-engineered n-MOSFETs.

10.5 Manufacturability Optimisation

So far, we only optimised device performance. Let us now add aspects of manufacturability, that is, the minimisation of the impact of parametric variations. We introduce the resulting variance in device characteristics as an optimisation constraint.

For each candidate from the performance optimisation, we evaluate the variance of device characteristics originating from the parametric variance.

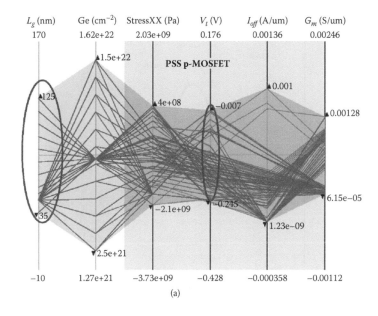

(a)

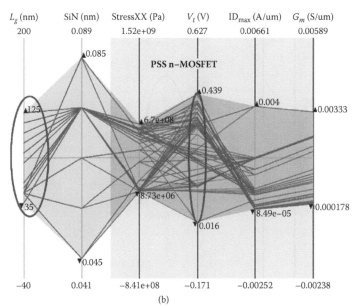

(b)

FIGURE 10.5
(a) Parallel coordinate plot. The process is optimised with respect to threshold voltage, current, and transconductance for process-induced strained Si p-MOSFETs. (b) Parallel coordinate plot. The process is optimised with respect to threshold voltage, current, and transconductance for process-induced strained Si n-MOSFETs.

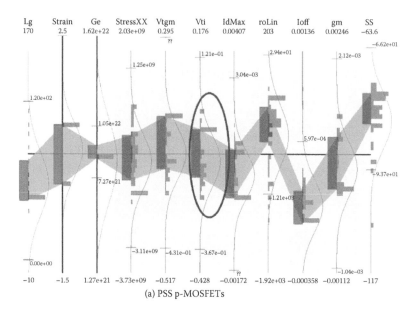

(a) PSS p-MOSFETs

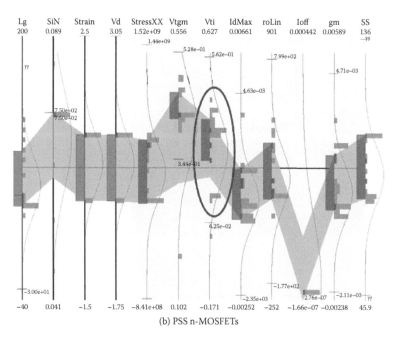

(b) PSS n-MOSFETs

FIGURE 10.6
(a) Parallel coordinate plot for process-induced strained Si p-MOSFETs created using PCM Studio, showing sensitivity analysis of design parameters and yield constraint variables. (b) Parallel coordinate plot for process-induced strained Si n-MOSFETs created using PCM Studio, showing sensitivity analysis of design parameters and yield constraint variables.

This is shown in Figure 10.6(a) and (b) for process-induced strained Si p- and n-MOSFETs. Input range constraints are as before; V_t is minimised. Scenarios show the optimised device with the variance of the output parameters. It can be seen that the variance of V_t is relatively high compared to the other device characteristics. It is shown that the final device performance is good and better from a manufacturability perspective.

The above DFM/PCM simulation example demonstrates how to optimise a process and reduce the process development time by reducing the number of costly and time-consuming design iterations.

10.6 Summary

With extreme scaling down of MOSFETs in high-volume manufacturing, it is imperative to develop a systematic TCAD-based methodology to design, characterise, and optimise manufacturability to increase yield. The process compact model has been used to find the optimum process conditions to meet a set of device specifications for strain-engineered MOSFETs. The interactive visual optimisation process using design of experiments in a parallel coordinate plot allows one to explore device performance criteria. Utilisation of TCAD tools for process optimisation for an overall design for manufacturing (DFM) solution is discussed.

Review Questions

1. What are the major sources of process variability?
2. What do you mean by intradie and interdie process variability? Give examples.
3. What are the patterning proximity effects?
4. What are the differences among (a) biaxial strain, (b) high-stress capping layers, and (c) embedded silicon-germanium (e-SiGe)?
5. How do high-k gate dielectric and a metal gate process affect the process variability?
6. What are design for manufacturability (DFM) and design for yield (DFY)?
7. Why does industry need yield-centric DFM?
8. Why is process optimisation needed?
9. Describe briefly: (a) sensitivity analysis, (b) uncertainty analysis, and (c) yield analysis.
10. What is the process window? Describe its importance.

References

1. S. Borkar, T. Karnik, S. Narendra, J. Tschanz, A. Keshavarzi, and V. De, Parameter Variations and Impacts on Circuits and Microarchitecture, *Proc. Design Automation Conference*, 338–342, 2003.
2. Semiconductor Industry Association, *International Technology Roadmap for Semiconductors (ITRS)*, 2009, http://public. itrs.net.
3. C. Ortolland, P. Morin, C. Chaton, E. Mastromatteo, C. Populaire, S. Orain, F. Leverd, P. Stolk, F. Boeuf, and F. Arnaud, Stress Memorization Technique (SMT) Optimization for 45 nm CMOS, *Proc. VLSI Tech. Dig.*, 78–79, 2006.
4. J. Jaffari and M. Anis, Variability-Aware Device Optimization under i_{ON} and Leakage Current Constraints, Proceedings of International Symposium on Low Power Electronics and Design (ISLPED 06), Tegernsee, Germany, October 2006.
5. C. C. Chiang and J. Kawa, *Design for Manufacturability and Yield for Nano-Scale CMOS*, Springer, Dordrecht, 2007.
6. M. Orshansky, S. R. Nassif, and D. Boning, *Design for Manufacturability and Statistical Design: A Constructive Approach*, Springer Science + Business Media, LLC, New York, USA, 2008.
7. A. Inselberg, The Plane with Parallel Coordinates, *Visual Computer*, 1, 69–91, 1985.
8. T. K. Maiti and C. K. Maiti, Strained-Engineered MOSFETs: Design and Optimization Using TCAD, *Proceedings of International SiGe Technology and Device Meeting (ISTDM)*, Taiwan, Mon-P1-60, 2008.

11

Conclusions

Strain engineering continues to evolve and will remain one of the key performance enablers for future generations of complementary metal-oxide-semiconductor (CMOS) technologies. In this monograph we have attempted to give some insight into process design, device modelling and simulation, and optimisation of strain-engineered metal-oxide-semiconductor field-effect transistors (MOSFETs). We have discussed the challenges involved in the heterogeneous integration of novel band-engineered materials and strained quantum wells on a Si platform. We have provided an overview of the major strain engineering techniques that have remarkably advanced the silicon CMOS transistor architecture, including embedded SiGe (e-SiGe), embedded SiC (e-SiC), the stress memorisation technique (SMT), dual-stress liners (DSLs), and the stress proximity technique (SPT). The application of local strain, however, is limited for further scaling beyond 22 nm, and as such, new methods of strain generation in the transistor channel region will be required. One possible option could be the combination of global and local strain. Multigate devices employing high-k gate dielectrics have emerged as a promising solution overcoming the scaling limitations of planar bulk CMOS. The advent of high-k/metal gate has brought additional strain benefits. Current techniques for generating strain in silicon are limited to a single type of strain (uniaxial or biaxial) within the substrate, which is clearly not optimum for simultaneous enhancement of hole and electron mobilities. Development of a new technology to generate biaxial and uniaxial silicon strain side by side on a single silicon-on-insulator (SOI) wafer will be of great technological importance.

Power dissipation is becoming a limiting factor in high-performance system design as technology scales and device integration level increases. Reduction in system power is not only important to improve battery life in portable devices, but it also plays an important role in enhancing the system reliability. Supply voltage scaling is emerging as an effective technique for reducing both dynamic and leakage power. Possible future innovations in device structures, novel process modules, and material systems have the potential to successfully address upcoming scaling challenges in a power-limited era. The convergence of new transistor structures and materials will be critical for successfully scaling CMOS transistors through the next decade. Strained Si is of technological interest for its ability to increase carrier mobilities in MOSFETs, and thereby improve circuit performance without requiring device scaling. At high vertical electric fields, biaxial tensile strain enhances electron mobility, while uniaxial compressive strain enhances hole mobility.

With the extreme scaling down of MOSFETs in high-volume manufacturing, it is imperative to develop technology computer-aided design (TCAD)-based methodology encompassing process design to device simulation, characterisation for SPICE parameter extraction, and optimisation of design for manufacturing to increase yield. A methodology for capturing process variability in SPICE models has been discussed. The model was validated by comparing device characteristics from the extracted SPICE parameters with those obtained from TCAD simulations. Variation-aware analogue and mixed-signal circuit design will require advanced models for process variations. We have focused on the process variation of strain-engineered MOSFETs at the device level, but circuit- and architecture-level approaches should be investigated to further mitigate process variations. The optimisation has been employed only on MOSFET design parameters, such as oxide thickness, gate length, channel doping concentration, etc. However, tape-out optimisation by modern lithography technologies is required. Furthermore, optimisation of all hierarchy levels (device, circuit, gate, and architecture) is required to suppress the process variation effects, reduce the power consumption, and improve the device performance.

We have discussed through-silicon via (TSV) modelling for simple structures. Optimisation of the size and placement of ground plugs to maximise noise isolation and to minimise area penalty is necessary. A comprehensive understanding of TSV stress-induced variation is also needed. Development of new techniques in mitigating TSV-induced substrate noise and their characterisation is also necessary. Toward understanding of reliability issues, negative bias temperature instability (NBTI) and hot-carrier injection (HCI) degradation in strain-engineered MOSFETs has been considered. The models used to study degradation mechanisms should further be modified for ultra-short-channel devices. Toward better understanding of the degradation mechanisms, such as circuit-level degradation, the combined effects of NBTI and HCI need to be considered. However, there will be an ultimate limit for the scaling when ballistic transport will take place. It is not clear if strain techniques will still be useful at that stage. Last, an integrated effort is necessary to work on novel CMOS structures involving (1) growth and fabrication of novel substrate materials, (2) device/circuit fabrication, (3) characterisation, and (4) modelling.

Index